ADVANCES IN INEQUALITIES FOR SERIES

ADVANCES IN INEQUALITIES FOR SERIES

SEVER S. DRAGOMIR AND ANTHONY SOFO
EDITORS

Nova Science Publishers, Inc.
New York

For permission to use material from this book please contact us:
Telephone 631-231-7269; Fax 631-231-8175
Web Site: http://www.novapublishers.com

Library of Congress Cataloging-in-Publication Data
Available upon request

ISBN-13: 978-1-60021-920-7
ISBN-10: 1-60021-920-9

Published by Nova Science Publishers, Inc. ✣ New York

Table of Contents

Preface

This research monograph, *Advances in Inequalities for Series*, deals with identities and inequalities relating to series and their application. This is the first volume of what we envisage to be a series of research monographs on advances in inequalities for series.

All of the papers in this volume have been fully peer reviewed. Some papers in this volume appear in print for the first time, detailing many technical results and some other papers offer a review of a number of recently published results.

The papers appear in author alphabetical order and not in mathematics subject classification. There are fifteen diverse papers in this volume each with its own speciality.

The first two papers of Babalola and Opoola investigate certain coefficient and radius problems for analytic functions of a particular subclass and satisfying a Salagean derivative operator inequality. Sharp bounds for the coefficients of this analytic function of a particular subclass, which is not necessarily univalent on the unit disk, are given.

An important issue in many applications of Probability theory is finding an approximate measure of distance, or discrimination, between two probability distributions. A number of divergence measure for this purpose have been proposed. The paper by Barnett, Cerone, Dragomir and Sofo, gives a survey of recent results as well as new approximations of f-Divergence measures via Taylor series expansions.

The paper by Berenhaut, Abernathy, Fan and Foley surveys several recent results related to inequalities for coefficients of reciprocals of power series under varying restrictions on the coefficients of the original series.

Burgin considers series in the context of hypernumbers. The theory of hypernumbers is a novel approach in functional analysis aimed at the development of mathematically correct techniques, which allows operations with integrals and series divergent in the classical sense and relations between inequalities.

The main result in the paper of Guney, is to prove coefficient inequalities and distortion theorems with the use of the Salagean differential operator on a subclass of analytic and p-valent functions with negative coefficients. In a similar vein Latha and Karthiyayini obtain coefficient estimates and extreme points on a class of multivalent functions. Continuing on this theme Latha and Shivarudrappa also obtain coefficient estimates for functions belonging to a subclass of analytic functions in a disk. Lei, Fang and Yang investigate and obtain some interesting results on the normality of meromorphic functions and shared sets. In the paper by Leindler, there is an investigation into equivalence theorems regarding almost decreasing numerical sequences and an application of estimations of new type of best approximation of functions of bounded variation by trigonometric polynomials.

Meng deals with the uniqueness problems on meromorphic functions concerning differential polynomials that share fixed points. Murugusundaramoorthy and Rosy investigate a sufficient condition, maximisation theorem and some inclusion relations associated with a new subclass of analytic functions of complex order, making use of the Salagean differential operator.

For a certain linear operator, which is defined by means of the Hadamard product with a generalised hypergeometric function, Shanmugam, Sivasubramanian and Darus investigate several mapping properties involving certain subclasses of analytic and univalent functions. Finally, Sofo investigates identities and inequalities involving reciprocals of binomial coefficients.

We trust that, in this research monograph, there is some topic that you find worthy of note, stimulating and perhaps sparks some idea for further research.

S. S. Dragomir and A. Sofo, Editors.

In: Advances in Inequalities for Series
Editors: S. S. Dragomir and A. Sofo, pp. 1–13
ISBN 978-1-60021-920-7

Chapter 1

On the Coefficients of a Certain Class of Analytic Functions

K. O. Babalola and T. O. Opoola
Department of Mathematics
University of Ilorin
Ilorin, Nigeria
E-mail addresses: abuqudduus@yahoo.com; opoola_stc@yahoo.com

Abstract

In this paper we solve certain coefficient problems for functions of the class $T_n^\alpha(\beta)$ introduced in [7] by Opoola. In particular, for some index $\alpha > 0$, we obtain sharp bounds for the coefficients of these functions. Furthermore we give some rough estimate of the general coefficient bounds using the logarithmic coefficient approach. We also prove the analogue of the well-known Fekete-Szego theorem for this class of functions.

1991 Mathematics Subject Classification: 30C45

Key words and phrases: Coefficient inequalities,Logarithmic coefficient, Salagean derivative operator, Analytic and univalent functions.

1. Introduction

Let A denote the class of functions:

$$f(z) = z + a_2 z^2 + \cdots \tag{1.1}$$

which are analytic in the unit disk $E = \{z : |z| < 1\}$. Also let P be the class of functions:

$$p(z) = 1 + p_1 z + p_2 z^2 + \cdots \tag{1.2}$$

which are analytic in E and have positive real part.

In [7] Opoola introduced the subclass $T_n^\alpha(\beta)$ consisting of functions $f \in A$ which satisfy the inequality:

$$\operatorname{Re} \frac{D^n f(z)^\alpha}{\alpha^n z^\alpha} > \beta, \tag{1.3}$$

where $\alpha > 0$ is real, $0 \leq \beta < 1$, D^n $(n \in N_0 = \{0,1,2,\ldots\})$ is the Salagean derivative operator defined as

$$D^n f(z) = D(D^{n-1}f(z)) = z[D^{n-1}f(z)]' \tag{1.4}$$

with $D^0 f(z) = f(z)$ and powers in (1.3) meaning principal values only. The geometric condition (1.3) slightly modifies the one given originally in [7] (see [2]).

The coefficient problem for various families of analytic and univalent functions is basically about the search for precise bounds of the coefficient a_k. This problem has attracted much attention from researchers in the last century. In this work we attempt to solve this problem for the class $T_n^{\alpha}(\beta)$. Our main tool is the close association which exists between the class $T_n^{\alpha}(\beta)$ and the family P of Caratheodory functions. Though the problem is not completely solved for all values the index α could assume, yet considerable breakthrough has been recorded in the determination of the *best possible* estimate for certain $\alpha > 0$. Furthermore we give some rough estimates of the general coefficient bounds using the logarithmic coefficient approach.

Another closely related problem is the determination of a sharp bound for the Fekete-Szego functional $|a_3 - \lambda a_2^2|$ for both real and complex values of the parameter λ. The Fekete-Szego problem has its origin in a conjecture of Littlewood and Parley (1932) that the true bound on the coefficients of an odd univalent function is 1. This was disproved in 1933 by Fekete and Szego via the determination of the sharp bound on the functional (see [3]). Thus the functional has continued to receive attention of researchers in geometric function theory.

We state some preliminary lemmas in Section 2. while our main results are proved in Section 3..

2. Preliminaries

We shall need the following well-known lemmas:

Lemma 2.1 ([5]). *If $|z|$ is less than the radius of convergence of either of the power series $\sum_{k=0} a_k z^k$ and $\sum_{k=0} b_k z^k$, then the product of the two series is $\sum_{k=0} c_k z^k$, where*

$$c_k = \sum_{j=0}^{k} a_j b_{k-j}. \tag{2.1}$$

Inductively, the following corollary is true.

Corollary 2.2. *Let $\sum_{k=0} c_k z^k$ be a power series. Then the mth integer product is*

$$\sum_{k=0} c_k^{(m)} z^k = \left(\sum_{k=0} c_k z^k\right)^m$$

where $c_k^{(1)} = c_k$ and

$$c_k^{(m)} = \sum_{j=0}^{k} c_j c_{k-j}^{(m-1)}, \qquad m = 2, 3, \ldots \tag{2.2}$$

Lemma 2.3 ([4]). *Let $p \in P$. Then $|p_k| \leq 2$, $k = 1,2,3\ldots$ Equality is attained for the Moebius function $L_0(z)$.*

Lemma 2.4 ([8]). *Let $p \in P$. Then*

$$\left| p_2 - \frac{p_1^2}{2} \right| \leq 2 - \frac{|p_1|^2}{2} \tag{2.3}$$

This result is sharp. Equality holds for the function

$$p(z) = \frac{1 + \frac{1}{2}(p_1 + \varepsilon \overline{p_1})z + \varepsilon z^2}{1 - \frac{1}{2}(p_1 - \varepsilon \overline{p_1})z - \varepsilon z^2}', \qquad |\varepsilon| = 1. \tag{2.4}$$

Corollary 2.5. *Let $p \in P$. Then we have the sharp inequalities*

(1) for any real number σ,

$$\left| p_2 - \sigma \frac{p_1^2}{2} \right| \leq \begin{cases} 2(1-\sigma), & if \sigma \leq 0; \\ 2, & if 0 \leq \sigma \leq 2; \\ 2(\sigma - 1), & if \sigma \geq 2. \end{cases} \tag{2.5}$$

(2) for any complex number σ,

$$\left| p_2 - \sigma \frac{p_1^2}{2} \right| \leq \begin{cases} 2, & if\, |1-\sigma| \leq 1; \\ 2|1-\sigma|, & if\, |1-\sigma| \geq 1. \end{cases} \tag{2.6}$$

Proof. We write

$$\left| p_2 - \sigma \frac{p_1^2}{2} \right| = \left| p_2 - \frac{p_1^2}{2} + \frac{p_1^2}{2} - \sigma \frac{p_1^2}{2} \right|. \tag{2.7}$$

By the triangle inequality and an application of Lemma 2.3, (2.7) gives

$$\left| p_2 - \sigma \frac{p_1^2}{2} \right| \leq 2 - \frac{|p_1^2|}{2}(1 - |1-\sigma|). \tag{2.8}$$

Now if $1 - |1-\sigma| \geq 0$ (that is $0 \leq \sigma \leq 2$) we have the second inequality of (2.5). Otherwise (that is $\sigma \leq 0$ or $\sigma \geq 2$) we have

$$\left| p_2 - \sigma \frac{p_1^2}{2} \right| \leq 2|1-\sigma|. \tag{2.9}$$

The inequality (2.9) gives the first inequality of (2.5) for $\sigma \leq 0$ and the third if $\sigma \geq 2$.

The inequalities in (2.6) follow easily from (2.8) in the case σ is a complex number. This completes the proof of the lemma. □

Lemma 2.6 ([3]). *Let $\varphi(z) = \sum_{k=1} \eta_k z^k$ be an arbitrary power series with positive radius of convergence and normalized by $\varphi(0) = 0$. Suppose $\exp(\varphi(z)) = \sum_{k=0} \mu_k z^k$, $\mu_0 = 1$. Then*

$$|\mu_k|^2 \leq \exp\left\{ \sum_{j=1}^{k} \left(j|\eta_j|^2 - \frac{1}{j} \right) \right\}.$$

Now the main results.

3. Main Results

Theorem 3.1. *Let $f \in T_n^\alpha(\beta)$. Define*

$$B_m = \frac{2^m(1-\beta)^m \alpha^{m(n-1)} \prod_{j=0}^{m-1}(1-j\alpha)}{m!}, \tag{3.1}$$

$$A_k(n,\alpha,\beta) = \sum_{m=1}^{k-1} B_m Q_{k-1}^{(m)}, \tag{3.2}$$

$$A_{k-1}(n,\alpha,\beta) = \sum_{m=1}^{k-2} B_m Q_{k-1}^{(m)}, \tag{3.3}$$

where for each $m = 1,2,\ldots$; $Q_{k-1}^{(m)}$ is defined by the power series

$$\left(\sum_{k=1} \frac{z^k}{(\alpha+k)^n}\right)^m = Q_m^{(m)} z^m + Q_{m+1}^{(m)} z^{m+1} + Q_{m+2}^{(m)} z^{m+2} + \cdots \tag{3.4}$$

Also let

$$\Omega_1 = \{\alpha | 0 < \alpha < (k-2)^{-1}, \quad k = 2,3,\ldots\},$$
$$\Omega_2 = \{\alpha | (k-2)^{-1} \leq \alpha \leq (k-3)^{-1}, \quad k = 4,6,\ldots\} \quad \text{and}$$
$$\Omega_3 = \{\alpha | (k-2)^{-1} \leq \alpha < (k-3)^{-1}, \quad k = 3,5,\ldots\}$$

Then

$$|a_k| \leq \begin{cases} A_k, & \text{if } \alpha \in \Omega_1 \cup \Omega_2; \\ A_{k-1}, & \text{if } \alpha \in \Omega_3. \end{cases} \tag{3.5}$$

The inequalities are sharp. Equalities are attained for $f(z)$ satisfying

$$\frac{D^n f(z)^\alpha}{\alpha^n z^\alpha} = \begin{cases} \frac{1+(1-2\beta)z}{1-z}, & \text{if } \alpha \in \Omega_1 \cup \Omega_2, \\ \frac{1+(1-2\beta)z^2}{1-z^2}, & \text{if } \alpha \in \Omega_3. \end{cases} \tag{3.6}$$

Proof. Since $f \in T_n^\alpha(\beta)$, there exists an analytic function $p_n(z)$ such that

$$f(z)^\alpha = z^\alpha[\beta + (1-\beta)p_n(z)], \tag{3.7}$$

where

$$p_n(z) = 1 + \sum_{k=1} \frac{\alpha^n}{(\alpha+k)^n} p_k z^k$$

(see [2, Lemma 4.2]). Thus we have

$$f(z) = z\left(1 + (1-\beta)\alpha^n \sum_{k=1} \frac{p_k z^k}{(\alpha+k)^n}\right)^{\frac{1}{\alpha}}. \tag{3.8}$$

Equation (3.8) expands binomially as

$$\frac{f(z)}{z} = 1 + \frac{(1-\beta)\alpha^n}{\alpha} \sum_{k=1} \frac{p_k z^k}{(\alpha+k)^n} + \frac{(1-\beta)^2\alpha^{2n}(1-\alpha)}{2!\alpha^2} \left(\sum_{k=1} \frac{p_k z^k}{(\alpha+k)^n} \right)^2 + \cdots$$
$$+ \frac{(1-\beta)^m \alpha^{mn} \prod_{j=0}^{m-1}(1-j\alpha)}{m!\alpha^m} \left(\sum_{k=1} \frac{p_k z^k}{(\alpha+k)^n} \right)^m + \cdots. \quad (3.9)$$

Using Corollary 2.2 in (3.9), we get

$$\frac{f(z)}{z} = 1 + \sum_{k=1} \tilde{B}_1 C_k^{(1)} z^k + \cdots + \sum_{k=m} \tilde{B}_m C_k^{(m)} z^k + \cdots, \quad (3.10)$$

where

$$\tilde{B}_m = \frac{(1-\beta)^m \alpha^{m(n-1)} \prod_{j=0}^{m-1}(1-j\alpha)}{m!},$$

$$C_k^{(m)} = \sum_{j=1}^{k-1} q_j C_{k-j}^{(m-1)}, \qquad m = 1, 2, 3, \ldots; \quad k = m, m+1, \ldots \quad (3.11)$$

and

$$q_k = \frac{p_k}{(\alpha+k)^n} \quad (3.12)$$

From (3.10) we have

$$f(z) = z + \sum_{k=2} \tilde{A}_k^{(m)} z^k, \quad (3.13)$$

where

$$\tilde{A}_k^{(m)} \equiv \tilde{A}_k^{(m)}(n,\alpha,\beta) = \sum_{m=1}^{k-1} \tilde{B}_m C_{k-1}^{(m)}. \quad (3.14)$$

Comparing coefficients in (1.1) and (3.13), we see that

$$a_k = \tilde{A}_k^{(m)}. \quad (3.15)$$

Now let $\alpha \in \Omega_1$, then the product $(1-\alpha)(1-2\alpha)\cdots(1-(m-1)\alpha)$ is nonnegative so that

$$\left|\tilde{A}_k^{(m)}\right| \leq \sum_{m=1}^{k-1} \tilde{B}_m \left|C_{k-1}^{(m)}\right|. \quad (3.16)$$

Observe from (3.11) and (3.12) that:

$$C_k^{(1)} = q_k, \quad C_k^{(k)} = q_1^k, \qquad k = 1, 2, 3, \ldots \quad (3.17)$$
$$C_3^{(2)} = 2q_1q_2;$$
$$C_4^{(2)} = q_2^2 + 2q_1q_3, \quad C_4^{(3)} = 3q_1^2q_2;$$
$$C_5^{(2)} = 2q_1q_4 + 2q_2q_3, \quad C_5^{(3)} = 3q_1q_2^2 + 3q_1^2q_3, \quad C_5^{(4)} = 4q_1^3q_2;$$

Thus, in general, $C_k^{(m)}$ has the form

$$C_k^{(m)} = \sum_{j=1}^{k} G_j \prod_{l=1}^{m} q_l^{\upsilon_l} \tag{3.18}$$

for some nonnegative constants G_j, $j = 1,2,\ldots,k$ and indices $\upsilon_l, l = 1,2,\ldots,m$ taking values in the set $M = \{0,1,2,\ldots,m\}$ such that the sum $\upsilon_1+\upsilon_2+\cdots+\upsilon_m = m$. Hence

$$\left|C_k^{(m)}\right| \le \sum_{j=1}^{k} G_j \prod_{l=1}^{m} |q_l^{\upsilon_l}| = \sum_{j=1}^{k} G_j \prod_{l=1}^{m} \left|\frac{p_l^{\upsilon_l}}{(\alpha+l)^{n\upsilon_l}}\right|. \tag{3.19}$$

Using Lemma 2.3, we obtain

$$\left|C_{k-1}^{(m)}\right| \le 2^m Q_{k-1}^{(m)}, \qquad m = 1,2,\ldots; \quad k = m+1, m+2, \ldots, \tag{3.20}$$

where

$$Q_{k-1}^{(m)} = \sum_{j=1}^{k-1} \frac{G_j}{\prod_{l=1}^{m} (\alpha+l)^{n\upsilon_l}} \tag{3.21}$$

is the coefficient defined by the power series (3.2). Therefore, equation (3.21) is equivalent to (3.18) where $q_k = 2/(\alpha+k)^n$. Using (3.20) in (3.15) and (3.16) we have the first inequality of (3.4) for $\alpha \in \Omega_1$. Now, suppose $\alpha \in \Omega_2 \cup \Omega_3$. We write (3.14) as

$$\tilde{A}_k^{(m)} = \tilde{A}_k^{(k-3)} + \tilde{B}_{k-2} C_{k-1}^{(k-2)} + \tilde{B}_{k-1} C_{k-1}^{(k-1)}. \tag{3.22}$$

That is,

$$\tilde{A}_k^{(m)} = \tilde{A}_k^{(k-3)} + \tilde{B}_{k-2}\left\{C_{k-1}^{(k-2)} + \frac{(1-\beta)\alpha^{(n-1)}(1-(k-2)\alpha)}{(k-1)} C_{k-1}^{(k-1)}\right\}. \tag{3.23}$$

Observe from (3.17) that the general form of $C_k^{(k-1)}$ and $C_k^{(k)}$ are respectively $C_k^{(k-1)} = (k-1)q_1^{k-2}q_2$ and $C_k^{(k)} = q_1^k$ so that, (3.23) becomes

$$\tilde{A}_k^{(m)} = \tilde{A}_k^{(k-3)} + (k-2)\tilde{B}_{k-2} q_1^{k-3}\left\{q_2 - \frac{2(1-\beta)\alpha^{(n-1)}((k-2)\alpha-1)}{(k-1)(k-2)}\frac{q_1^2}{2}\right\}. \tag{3.24}$$

Using (3.11) in (3.24), we obtain

$$\tilde{A}_k^{(m)} = \tilde{A}_k^{(k-3)} + \frac{(k-2)\tilde{B}_{k-2}}{(\alpha+1)^{n(k-3)}(\alpha+2)^n} p_1^{k-3}\left\{p_2 - \sigma\frac{p_1^2}{2}\right\}, \tag{3.25}$$

where

$$\sigma = \frac{2(1-\beta)\alpha^{(n-1)}((k-2)\alpha-1)(\alpha+2)^n}{(k-1)(k-2)(\alpha+1)^{2n}}$$

is a real number between 0 and 1. Hence by Corollary 2.5, the absolute value of the term in the Poison brackets in (3.25) attains a maximum value 2. Therefore, from (3.25) we have

$$\left|\tilde{A}_k^{(m)}\right| \le \left|\tilde{A}_k^{(k-3)}\right| + \frac{2^{k-2}(k-2)\tilde{B}_{k-2}}{(\alpha+1)^{n(k-3)}(\alpha+2)^n}. \tag{3.26}$$

That is

$$\left|\tilde{A}_k^{(m)}\right| \leq \left|\tilde{A}_k^{(k-3)}\right| + B_{k-2}Q_{k-1}^{(k-2)}, \tag{3.27}$$

where

$$Q_{k-1}^{(k-2)} = \frac{k-2}{(\alpha+1)^{n(k-3)}(\alpha+2)^n}.$$

By (3.15), (3.16) and (3.20), we have from (3.27)

$$|a_k| \leq A_{k-1} \tag{3.28}$$

for $\alpha \in \Omega_2 \cup \Omega_3$. From (2.4), we find that the boundary functions are given by

$$\frac{D^n f(z)^\alpha}{\alpha^n z^\alpha} = \frac{1+(1-2\beta)z^2}{1-z^2}. \tag{3.29}$$

However, with regard to the even coefficients, we notice a disagreement between the upper bound in (3.28) and the coefficients in (3.29). This seeming contradiction suggests a possibility of refinement for the bound on the even coefficients. This is achieved thus: we write the inequality (2.3) of Lemma 2.4 as

$$p_2 = \tfrac{1}{2}p_1^2 + \varepsilon\left(2 - \tfrac{1}{2}|p_1|^2\right), \qquad |\varepsilon| \leq 1. \tag{3.30}$$

By (3.30), we eliminate p_2 in (3.25) and we find that

$$\tilde{A}_k^{(m)} = \tilde{A}_k^{(k-3)} + \frac{(k-2)\tilde{B}_{k-2}}{(\alpha+1)^{n(k-3)}(\alpha+2)^n} p_1^{k-3}\left\{(1-\sigma)\frac{p_1^2}{2} + \varepsilon\left(2 - \tfrac{1}{2}|p_1|^2\right)\right\}. \tag{3.31}$$

Since $\left(2 - \frac{1}{2}|p_1|^2\right) \geq 0$, the absolute value of the term in the Poisson brackets in (3.31) attains its maximum for $|p_1| = 2$. Hence, for $\alpha \in \Omega_2$we have from (3.31)

$$\left|\tilde{A}_k^{(m)}\right| \leq \left|\tilde{A}_k^{(k-3)}\right| + \frac{2^{k-2}(1-\sigma)(k-2)\tilde{B}_{k-2}}{(\alpha+1)^{n(k-3)}(\alpha+2)^n}. \tag{3.32}$$

The inequality (3.32) together with (3.15), (3.16) and (3.20) and the fact that

$$Q_{k-1}^{(k-2)} = \frac{k-2}{(\alpha+1)^{n(k-3)}(\alpha+2)^n} \quad \text{and} \quad Q_{k-1}^{(k-1)} = \frac{1}{(\alpha+1)^{n(k-1)}}$$

gives

$$|a_k| \leq A_k \tag{3.33}$$

for $\alpha \in \Omega_2$. This completes the proof . □

Remark 3.1.

(1) *If we set $n = \beta = 0$, $\alpha = 1$, we immediately obtain the sharp inequality $|a_k| \leq 2$, $k = 2, 3, \ldots$, which was reported in [11].*

(2) *Putting $n = \alpha = 1$, $\beta = 0$, we get a result of Macgregor [6, Theorem 1(1)].*

(3) *By making $n = 1$ and $\beta = 0$, we obtain (in part) a result of Singh [10, Theorem 6].*

(4) *We observe from (3.4) that:*

$$Q_1^{(1)} = \frac{1}{(\alpha+1)^n};$$

$$Q_2^{(1)} = \frac{1}{(\alpha+2)^n}, \qquad Q_2^{(2)} = \frac{1}{(\alpha+1)^{2n}};$$

$$Q_3^{(1)} = \frac{1}{(\alpha+3)^n}, \qquad Q_3^{(2)} = \frac{2}{(\alpha+1)^n(\alpha+2)^n}, \qquad Q_3^{(3)} = \frac{1}{(\alpha+1)^{3n}};$$

$$\begin{aligned} Q_4^{(1)} &= \frac{1}{(\alpha+4)^n}, \\ Q_4^{(2)} &= \frac{1}{(\alpha+2)^{2n}} + \frac{2}{(\alpha+1)^n(\alpha+3)^n}, \\ Q_4^{(3)} &= \frac{3}{(\alpha+1)^{2n}(\alpha+2)^n}, \\ Q_4^{(4)} &= \frac{1}{(\alpha+1)^{4n}}; \end{aligned}$$

$$\begin{aligned} Q_5^{(1)} &= \frac{1}{(\alpha+5)^n}, \\ Q_5^{(2)} &= \frac{2}{(\alpha+1)^n(\alpha+4)^n} + \frac{2}{(\alpha+2)^n(\alpha+3)^n}, \\ Q_5^{(3)} &= \frac{3}{(\alpha+1)^n(\alpha+2)^{2n}} + \frac{3}{(\alpha+1)^{2n}(\alpha+3)^n}, \\ Q_5^{(4)} &= \frac{4}{(\alpha+1)^{3n}(\alpha+2)^n}, \\ Q_5^{(5)} &= \frac{1}{(\alpha+1)^{5n}}, \qquad \ldots \end{aligned}$$

(5) *Using (4) above in (3.1), (3.2) and (3.3), we can write out some of the bounds in explicit terms as follows:*

$$|a_2| \leq \frac{2(1-\beta)\alpha^{n-1}}{(\alpha+1)^n}, \qquad \alpha > 0$$

$$|a_3| \leq \begin{cases} \frac{2(1-\beta)\alpha^{n-1}}{(\alpha+2)^n} + \frac{2(1-\beta)^2(1-\alpha)\alpha^{2n-2}}{(\alpha+1)^{2n}}, & 0 < \alpha \leq 1; \\ \frac{2(1-\beta)\alpha^{n-1}}{(\alpha+2)^n}, & \alpha \geq 1, \end{cases}$$

$$|a_4| \leq \frac{2(1-\beta)\alpha^{n-1}}{(\alpha+3)^n} + \frac{4(1-\beta)^2(1-\alpha)\alpha^{2n-2}}{(\alpha+1)^n(\alpha+2)^n} + \frac{4(1-\beta)^3(1-\alpha)(1-2\alpha)\alpha^{3n-3}}{3(\alpha+1)^{3n}}, \qquad 0 < \alpha \leq 1;$$

$$|a_5| \le \frac{2(1-\beta)\alpha^{n-1}}{(\alpha+2)^n} + \frac{2(1-\beta)^2(1-\alpha)[2(\alpha+2)^{2n}+(\alpha+1)^n(\alpha+3)^n]\alpha^{2n-2}}{(\alpha+1)^n(\alpha+2)^{2n}(\alpha+3)^n}$$
$$+\frac{4(1-\beta)^3(1-\alpha)(1-2\alpha)\alpha^{3n-3}}{(\alpha+1)^{2n}(\alpha+2)^n} + \frac{2(1-\beta)^4(1-\alpha)(1-2\alpha)(1-3\alpha)\alpha^{4n-4}}{3(\alpha+1)^{4n}}$$

if $0 < \alpha \le \frac{1}{3}$;

$$|a_5| \le \frac{2(1-\beta)\alpha^{n-1}}{(\alpha+2)^n} + \frac{2(1-\beta)^2(1-\alpha)[2(\alpha+2)^{2n}+(\alpha+1)^n(\alpha+3)^n]\alpha^{2n-2}}{(\alpha+1)^n(\alpha+2)^{2n}(\alpha+3)^n}$$
$$+\frac{4(1-\beta)^3(1-\alpha)(1-2\alpha)\alpha^{3n-3}}{(\alpha+1)^{2n}(\alpha+2)^n},$$

if $\frac{1}{3} \le \alpha \le \frac{1}{2}$;

$$|a_6| \le \frac{2(1-\beta)\alpha^{n-1}}{(\alpha+5)^n} + \frac{4(1-\beta)^2(1-\alpha)[(\alpha+1)^n(\alpha+4)^n+(\alpha+2)^n(\alpha+3)^n]\alpha^{2n-2}}{(\alpha+1)^n(\alpha+2)^n(\alpha+3)^n(\alpha+4)^n}$$
$$+\frac{4(1-\beta)^3(1-\alpha)(1-2\alpha)[(\alpha+1)^n(\alpha+3)^n+(\alpha+2)^{2n}]\alpha^{3n-3}}{(\alpha+1)^{2n}(\alpha+2)^{2n}(\alpha+3)^n}$$
$$+\frac{8(1-\beta)^4(1-\alpha)(1-2\alpha)(1-3\alpha)\alpha^{4n-4}}{3(\alpha+1)^{3n}(\alpha+2)^n}$$
$$+\frac{4(1-\beta)^5(1-\alpha)(1-2\alpha)(1-3\alpha)(1-4\alpha)\alpha^{5n-5}}{15(\alpha+1)^{5n}}, \qquad \left(0<\alpha\le\frac{1}{3}\right)$$

and so on.

Let S denote the subclass of A consisting only of univalent functions in E. Associated with each function $f \in S$ is a well-defined logarithmic function

$$\log\frac{f(z)}{z} = 2\sum_{k=1}\gamma_k z^k, \qquad z\in E.$$

The numbers γ_k are called the logarithmic coefficients of $f(z)$ (see [3]). The relevance of the logarithmic coefficients comes from the fact that, by the so called Lebedev-Milin inequality (Lemma 2.6), estimates of the logarithmic coefficients of $f(z)$ can be transferred to bounds on the coefficients of $f(z)$.

In the following theorem we shall associate with each $f \in T_n^\alpha(\beta)$ logarithmic coefficients:

$$\log\frac{f(z)^\alpha}{z^\alpha} = 2\alpha\sum_{k=1}\gamma_k z^k, \qquad z\in E \tag{3.34}$$

and use it to approximate the general bounds on the coefficients of $f(z)$.

Theorem 3.2. *Let* $f \in T_n^\alpha(\beta)$, $n\in\mathbb{N}$. *Suppose*

$$\frac{f(z)^\alpha}{z^\alpha} = \sum_{k=0} A_{k+1}(\alpha)z^k, \qquad A_1(\alpha)=1, \tag{3.35}$$

then

$$|A_{k+1}(\alpha)| < \exp\left\{0.624\alpha^2 + \left(2\alpha^2-\tfrac{1}{2}\right)\sum_{j=1}^{k}\frac{1}{j}\right\}. \tag{3.36}$$

Proof. From (3.34) and (3.35) we have

$$\log\left\{\sum_{k=0} A_{k+1}(\alpha)z^k\right\} = \sum_{k=1} 2\alpha\gamma_k z^k,$$

that is,

$$\sum_{k=0} A_{k+1}(\alpha)z^k = \exp\left\{\sum_{k=1} 2\alpha\gamma_k z^k\right\}. \tag{3.37}$$

By Lemma 2.6, we get

$$|A_{k+1}(\alpha)|^2 \le \exp\left\{\sum_{j=1}^{k}\left(4\alpha^2 j|\gamma_j|^2 - \frac{1}{j}\right)\right\}. \tag{3.38}$$

Hence by Milin's lemma (for each $f \in S$)

$$\sum_{j=1}^{k} j|\gamma_j|^2 \le \sum_{j=1}^{k}\frac{1}{j} + \delta, \qquad k = 1, 2, \ldots.$$

where $\delta < 0.312$ is called the *Milin's constant*, (see [3])), the inequality (3.38) yields

$$|A_{k+1}(\alpha)|^2 \le \exp\left\{4\alpha^2\delta + (4\alpha^2 - 1)\sum_{j=1}^{k}\frac{1}{j}\right\}, \tag{3.39}$$

which implies (3.36) and thus completes the proof. □

Remark 3.2. *If* $\alpha = \frac{1}{2}$, *then* $\left|A_{k+1}\left(\frac{1}{2}\right)\right| < 1.168826$, $k = 1, 2, \ldots$ *In particular* $\left|A_2\left(\frac{1}{2}\right)\right| = \frac{1}{2}|a_2|$ *and* $\left|A_3\left(\frac{1}{2}\right)\right| = \left|\frac{1}{2}a_3 - \frac{1}{2}\left(\frac{1}{2}a_2\right)^2\right|$. *Thus, we have* $|a_2| < 2.337652$ *and* $|a_3| < 3.703806$. *Obviously these estimates are not sharp. However this approach can conveniently lead to a rough estimate of the bounds on the coefficients of functions in the class.*

Theorem 3.3. *Let* $f \in T_n^{\alpha}(\beta)$. *Then*

(i) for any real number λ,

$$|a_3 - \lambda a_2^2| \le \begin{cases} \frac{2(1-\beta)\alpha^{n-1}}{(\alpha+2)^n} - \frac{2(2\lambda+\alpha-1)(1-\beta)^2\alpha^{2n-2}}{(\alpha+1)^{2n}}, & \text{if } \lambda \le \frac{1-\alpha}{2} \\ \frac{2(1-\beta)\alpha^{n-1}}{(\alpha+2)^n}, & \text{if } \frac{1-\alpha}{2} \le \lambda \le \frac{1-\alpha}{2} + \frac{\alpha^{1-n}(\alpha+1)^{2n}}{(1-\beta)(\alpha+2)^n} \\ \frac{2(2\lambda+\alpha-1)(1-\beta)^2\alpha^{2n-2}}{(\alpha+1)^{2n}} - \frac{2(1-\beta)\alpha^{n-1}}{(\alpha+2)^n}, & \text{if } \lambda \ge \frac{1-\alpha}{2} + \frac{\alpha^{1-n}(\alpha+1)^{2n}}{(1-\beta)(\alpha+2)^n} \end{cases}$$

(ii) for any complex number λ,

$$|a_3-\lambda a_2^2| \le \begin{cases} \frac{2(1-\beta)\alpha^{n-1}}{(\alpha+2)^n}, & if \quad \left|1-\frac{(2\lambda+\alpha-1)(1-\beta)(\alpha+2)^n\alpha^{n-1}}{(\alpha+1)^{2n}}\right| \le 1 \\ \frac{2(1-\beta)\alpha^{n-1}|(\alpha+1)^{2n}-(2\lambda+\alpha-1)(1-\beta)^2\alpha^{2n-2}|}{(\alpha+1)^{2n}(\alpha+2)^n}, & \\ & if \quad \left|1-\frac{(2\lambda+\alpha-1)(1-\beta)(\alpha+2)^n\alpha^{n-1}}{(\alpha+1)^{2n}}\right| \ge 1. \end{cases}$$

The inequalities are sharp. For each λ*, there is a function* $f \in T_n^\alpha(\beta)$ *for which equality holds.*

Proof. From equations (3.14), (3.15) and (3.16) of Theorem 3.1 we get

$$a_2 = \frac{(1-\beta)\alpha^{n-1}}{(\alpha+1)^n} p_1$$

and

$$a_3 = \frac{(1-\beta)\alpha^{n-1}}{(\alpha+2)^n} p_2 + \frac{(1-\alpha)(1-\beta)^2\alpha^{2n-2}}{2(\alpha+1)^{2n}} p_1^2.$$

Hence

$$\left|a_3-\lambda a_2^2\right| = \frac{(1-\beta)\alpha^{n-1}}{(\alpha+2)^n}\left|p_2 - \frac{(2\lambda+\alpha-1)(1-\beta)(\alpha+2)^n\alpha^{n-1}}{(\alpha+1)^{2n}}\frac{p_1^2}{2}\right|.$$

By choosing

$$\sigma = \frac{(2\lambda+\alpha-1)(1-\beta)(\alpha+2)^n\alpha^{n-1}}{(\alpha+1)^{2n}}$$

in Corollary 2.5, the theorem follows. □

Remark 3.3. *Putting* $n=1$ *and* $\beta=0$*, we deduce a result of Singh [10, Theorem 7].*

Bibliography

[1] S. Abdulhalim, On a class of analytic functions involving the Salagean differential operator, *Tamkang J. Maths.,* **23**(1) (1992), 51–58.

[2] K.O. Babalola and T.O. Opoola, Iterated integral transforms of Caratheodory functions and their applications to analytic and univalent functions, *to appear*.

[3] P.L. Duren, Coefficients of univalent functions. *Bull. Amer. Math. Soc.,* **83** (1977), 891–911.

[4] P.L. Duren, *Univalent functions.* Springer-Verlag, New York, Inc. 1983

[5] G.H. Hardy, *A First Course in Pure Mathematics,* Cambridge University Press, 10th Edition, 1952.

[6] T.H. MacGregor, Function whose derivatives have positive real part, *Trans. Amer. Math. Soc.,* **104** (1962), 532 -537. MR 25 – 797

[7] T.O. Opoola, On a new subclass of univalent function. *Matematika* (36) 59 No. 2 (1994), 195–200.

[8] Ch. Pommerenke, *Univalent Functions,* Springer-Verlag. New York Inc. 1983.

[9] G.S. Salagean, Subclasses of univalent functions, *Lecture Notes in Math.,* 1013 (1983), 362–372. Springer-Verlag, Berlin, Heidelberg & New York.

[10] R. Singh, On Bazilevic functions, *Proc. Amer. Math. Soc.,* **38** (1973), 261–271.

[11] K. Yamaguchi, On functions satisfying Re $\{f(z)/z\} > 0$, *Proc. Amer. Math. Soc.,* **17** (1966), 588–591.

In: Advances in Inequalities for Series
Editors: S. S. Dragomir and A. Sofo, pp. 15–21
ISBN 978-1-60021-920-7

Chapter 2

Radius Problems for a Certain Class of Analytic Functions Involving the Salagean Derivative

K. O. Babalola and T. O. Opoola
Department of Mathematics
University of Ilorin
Ilorin, Nigeria
E-mail addresses: abuqudduus@yahoo.com; opoola_stc@yahoo.com

Abstract

In this paper we determine the radii of starlikeness and convexity of functions of the class $T_n^{\alpha}(\beta)$ introduced in [3] by Opoola. It is well known that for $n = 0$, this class consists largely of analytic functions, which are not necessarily univalent in the unit disk. For this case $n = 0$, the radius of univalence is indicated by the sufficiency of starlikeness.

1991 Mathematics Subject Classification: 30C45

Key words and phrases: Radius problem, Salagean derivative operator, Analytic and univalent functions.

1. Introduction

Let A denote the class of functions:

$$f(z) = z + a_2 z^2 + \cdots \tag{1.1}$$

which are analytic in the unit disk $E = \{z : |z| < 1\}$. Also let P be the class of functions:

$$p(z) = 1 + p_1 z + p_2 z^2 + \cdots \tag{1.2}$$

which are analytic in E and have positive real part.

In [3] Opoola introduced the subclass $T_n^\alpha(\beta)$ consisting of functions $f \in T$ which satisfy the inequality:

$$\mathrm{Re}\,\frac{D^n f(z)^\alpha}{\alpha^n z^\alpha} > \beta \tag{1.3}$$

where $\alpha > 0$ is real, $0 \le \beta < 1$, D^n $(n \in N_0 = \{0,1,2,\ldots\})$ is the Salagean derivative operator defined as

$$D^n f(z) = D(D^{n-1} f(z)) = z[D^{n-1} f(z)]' \tag{1.4}$$

with $D^0 f(z) = f(z)$ and powers in (1.3) meaning principal values only. The geometric condition (1.3) slightly modifies the one given originally in [3] (see [1]).

It is well-known that for $f \in A$ the condition that

$$\mathrm{Re}\,\frac{zf'(z)}{f(z)} > 0, \qquad |z| < r \tag{1.5}$$

is necessary and sufficient for starlikeness (and univalence) in the disk $|z| < r$, that is the function $f(z)$ maps the disk $|z| < r$ onto a starlike domain. Also a necessary and sufficient condition for $f \in A$ to be convex in the disk $|z| < r$ is that

$$\mathrm{Re}\left(1 + \frac{zf''(z)}{f'(z)}\right) > 0, \qquad |z| < r \tag{1.6}$$

If we know that a function maps the unit disk onto a domain having some nice geometric properties such as above, then we have the means of a more penetrating study of the function. These classes of functions were generalized by G.S. Salagean in [4]. Salagean stated that a function $f \in A$ belongs to the class $S_n(\lambda)$, $0 \le \lambda < 1$ if and only if

$$\mathrm{Re}\,\frac{D^{n+1} f(z)}{D^n f(z)} > \lambda, \qquad |z| < r \tag{1.7}$$

where for $n = \lambda = 0$ we have (1.5) and $n = 1$, $\lambda = 0$ gives (1.6).

In the present paper we solve the conventional two radius problems for the class $T_n^\alpha(\beta)$. Our approach is to determine, for functions of this class, the radius of the largest disk $E_0 = \{z : |z| < r_0(\alpha,\beta)\}$ in which the conditions (1.7) holds for $\lambda = 0$. Thus the radii of starlikeness and convexity of functions in $T_n^\alpha(\beta)$ follows as simple corollaries. The class $T_n^\alpha(\beta)$ is a generalization of many known classes of function (see [3]). Thus the result of this work will naturally lead to those known about such classes. In particular, our main result generalizes those of Macgregor [2], Tuan and Anh [5] and Yamaguchi [6].

The main result of our paper is the following:

Theorem 1.1. *Let $f \in T_n^\alpha(\beta)$. Then $f \in S_n(0)$ for $|z| < r_0(\alpha,\beta)$ where*

$$r_0(\alpha,\beta) = \begin{cases} \frac{[(1+\alpha^2)(1-\beta)^2 - 2\alpha\beta(1-\beta)]^{\frac{1}{2}} + \beta(1+\alpha) - 1}{\alpha(1-2\beta)} & \text{if } 0 \le \beta \le \beta_0(\alpha) \\ \left(\frac{4[\alpha\beta(1-\beta)]^{\frac{1}{2}} - [(1-\alpha^2)(1-\beta) + 4\alpha\beta]}{(1+\alpha^2)(1-\beta) + 2\alpha(1-3\beta)}\right)^{\frac{1}{2}} & \text{if } \beta_0(\alpha) \le \beta < 1,\ \beta \ne \frac{(1+\alpha)^2}{1+6\alpha+\alpha^2} \\ \left(\frac{\alpha}{1+\alpha}\right)^{\frac{1}{2}} & \text{if } \beta = \frac{(1+\alpha)^2}{1+6\alpha+\alpha^2}. \end{cases} \tag{1.8}$$

The results are the best possible.

Corollary 1.2. *Each function $f \in T_0^{\alpha}(\beta)$ maps the disk $|z| < r_0(\alpha,\beta)$ onto a starlike domain, where $r_0(\alpha,\beta)$ is defined by (1.8).*

The results are the best possible.

Corollary 1.3. *Each function $f \in T_1^{\alpha}(\beta)$ maps the disk $|z| < r_0(\alpha,\beta)$ onto a convex domain, where $r_0(\alpha,\beta)$ is defined by (1.8).*

The results are the best possible.

In Section 2., we state some preliminary lemmas, which we will depend on in the proof of our result in Section 3..

2. Preliminary Lemmas

We shall need the following lemmas:

Lemma 2.1 ([5]). *Let $p \in P$. Then for $0 \le \beta < 1$ and $|z| = r < 1$,*

$$\mathrm{Re}\left\{\frac{zp'(z)}{\frac{\beta}{1-\beta}+p(z)}\right\} \ge \begin{cases} \frac{-2(1-\beta)r}{(1+r)(1+(2\beta-1)r)}, & \text{for } R_1 \le R_2, \\ \frac{-\beta}{1-\beta}+\frac{1}{1-\beta}\left[2R_1 - \frac{1-(2\beta-1)r^2}{1-r^2}\right] & \text{for } R_2 \le R_1, \end{cases} \tag{2.1}$$

where

$$R_1 = \left(\frac{\beta-\beta(2\beta-1)r^2}{1-r^2}\right)^{\frac{1}{2}} \quad \text{and} \quad R_2 = \frac{1+(2\beta-1)r}{1+r} \tag{2.2}$$

The functions given by

$$p(z) = \begin{cases} \frac{1-z}{1+z} & \text{for } R_1 \le R_2, \\ \frac{1+z}{1-z} & \text{for } R_2 \le R_1, \end{cases} \tag{2.3}$$

show that the inequalities are sharp.

For a given $r < 1$, the transition from the first case $R_1 \le R_2$ to the second case $R_2 \le R_1$ takes place when $\beta = \beta_0 \in (0,1)$, where β_0 is determined by the equation $R_1 = R_2$.

Lemma 2.2. *Let $f \in A$, and $\alpha > 0$ be real. If $D^{n+1}f(z)^{\alpha}/D^n f(z)^{\alpha}$ takes a value which is independent of n, then*

$$\frac{D^{n+1}f(z)^{\alpha}}{D^n f(z)^{\alpha}} = \alpha\frac{D^{n+1}f(z)}{D^n f(z)}. \tag{2.4}$$

Proof. Let $D^{n+1}f(z)^{\alpha}/D^n f(z)^{\alpha}$ assume the same value for all $n \in N_0$. For $n = 0$, the assertion is easy to verify. Let $n = 1$. Then

$$\frac{D^2 f(z)^{\alpha}}{D^1 f(z)^{\alpha}} = 1 + \frac{zf''}{f'} + (\alpha-1)\frac{zf'}{f} = \frac{D^2 f(z)}{D^1 f(z)} + \left(1-\frac{1}{\alpha}\right)\frac{D^1 f(z)^{\alpha}}{D^0 f(z)^{\alpha}}.$$

But $D^1 f(z)^\alpha / D^0 f(z)^\alpha = D^2 f(z)^\alpha / D^1 f(z)^\alpha$ so that

$$\frac{D^2 f(z)^\alpha}{D^1 f(z)^\alpha} = \alpha \frac{D^2 f(z)}{D^1 f(z)}.$$

Next assume that (2.4) is true for some integer k. Then

$$\frac{D^{k+2} f(z)^\alpha}{D^{k+1} f(z)^\alpha} = (\alpha - 1) \frac{D^{k+1} f(z)}{D^k f(z)} + \frac{D^{k+2} f(z)}{D^{k+1} f(z)}$$

and

$$\frac{D^{k+2} f(z)^\alpha}{D^{k+1} f(z)^\alpha} = \left(1 - \frac{1}{\alpha}\right) \frac{D^{k+1} f(z)^\alpha}{D^k f(z)^\alpha} + \frac{D^{k+2} f(z)}{D^{k+1} f(z)}. \tag{2.5}$$

But $D^{k+1} f(z)^\alpha / D^k f(z)^\alpha$ has the same value for each k so that (2.5) can be written as

$$\frac{D^{k+2} f(z)^\alpha}{D^{k+1} f(z)^\alpha} = \left(1 - \frac{1}{\alpha}\right) \frac{D^{k+2} f(z)^\alpha}{D^{k+1} f(z)^\alpha} + \frac{D^{k+2} f(z)}{D^{k+1} f(z)},$$

that is,

$$\frac{D^{k+2} f(z)^\alpha}{D^{k+1} f(z)^\alpha} = \alpha \frac{D^{k+2} f(z)}{D^{k+1} f(z)}. \tag{2.6}$$

Hence the lemma follows by induction. □

We now turn to the proof of our main result.

3. Proof of the Main Result

Let $f \in T_n^\alpha(\beta)$, then there exists $p \in P$ such that

$$D^n f(z)^\alpha = \alpha^n z^\alpha [\beta + (1 - \beta) p(z)]. \tag{3.1}$$

Logarithmic differentiation of (3.1) yields

$$\frac{D^{n+1} f(z)^\alpha}{D^n f(z)^\alpha} = \alpha + \frac{z p'(z)}{\frac{\beta}{1-\beta} + p(z)}. \tag{3.2}$$

Since the right hand of (3.2) is independent of n, it follows by Lemma 2.2 that

$$\alpha \frac{D^{n+1} f(z)}{D^n f(z)} = \alpha + \frac{z p'(z)}{\frac{\beta}{1-\beta} + p(z)}. \tag{3.3}$$

Now let R_1 and R_2 be as in Lemma 2.1 and suppose $R_1 \leq R_2$. Then applying the lemma we have

$$\alpha \operatorname{Re} \frac{D^{n+1} f(z)}{D^n f(z)} \geq \alpha - \frac{2(1-\beta) r}{(1+r)(1+(2\beta-1) r)}. \tag{3.4}$$

It is easy to deduce from (3.4) that

$$\operatorname{Re} \frac{D^{n+1} f(z)}{D^n f(z)} > 0, \tag{3.5}$$

provided $|z| = r < r_0(\alpha,\beta)$, where $r_0(\alpha,\beta)$ is the smallest positive root of the equation

$$\alpha(2\beta-1)r^2+2(\alpha\beta+\beta-1)r+\alpha=0, \tag{3.6}$$

which is the first part of (1.8).

Evidently $r_0(\alpha,\beta)$, in this case, is not real except $\beta \leq (1+\alpha^2)/(1+\alpha)^2$. Furthermore since $r_0(\alpha,\beta) < 1$, we must have $\beta < 1/(1+2\alpha)$.

Next we suppose $R_2 \leq R_1$ then using Lemma 2.1 again, we have

$$\alpha \operatorname{Re} \frac{D^{n+1}f(z)}{D^n f(z)} \geq \alpha - \frac{\beta}{1-\beta} + \frac{1}{1-\beta}\left[2R_1 - \frac{1-(2\beta-1)r^2}{1-r^2}\right], \tag{3.7}$$

which implies that the condition (3.5) is satisfied if $|z| = r < r_0(\alpha,\beta)$, where $r_0(\alpha,\beta)$ is the smallest positive root of the equation

$$ar^4+br^2+c=0, \tag{3.8}$$

where

$$\begin{aligned} a &= (1-\beta)^2(1+\alpha^2)+2\alpha(1-\beta)(1-3\beta), \\ b &= 2(1-\beta)^2(1-\alpha^2)+8\alpha\beta(1-\beta), \\ c &= \alpha\beta(\alpha\beta+2\beta-2\alpha)+(1-\beta)^2+\alpha^2-2\alpha. \end{aligned}$$

If $\beta \neq (1+\alpha)^2/(1+6\alpha+\alpha^2)$, the solution of (3.8) is the second part of (1.8), otherwise we have the last segment of (1.8).

From equation (2.3) we deduce that the functions are given by

$$\frac{D^n f(z)^\alpha}{\alpha^n z^\alpha} = \begin{cases} \frac{1-(1-2\beta)z}{1+z}, & \text{if} \quad 0 \leq \beta \leq \beta_0(\alpha), \\ \frac{1+(1-2\beta)z}{1-z}, & \text{if} \quad \beta_0(\alpha) \leq \beta < 1, \end{cases} \tag{3.9}$$

show that the results are the best possible.

For any α, the number $\beta_0(\alpha) \in (0,1)$ is the solution of the equation:

$$\left(\frac{4[\alpha\beta(1-\beta)]^{\frac{1}{2}}-[(1-\alpha^2)(1-\beta)+4\alpha\beta]}{(1+\alpha^2)(1-\beta)+2\alpha(1-3\beta)}\right)^{\frac{1}{2}} = \frac{[(1+\alpha^2)(1-\beta)^2-2\alpha\beta(1-\beta)]^{\frac{1}{2}}+\beta(1+\alpha)-1}{\alpha(1-2\beta)}. \tag{3.10}$$

For instance, if $\alpha = 1$, we have $\beta_0(1) = \frac{1}{10}$ as in [5, Theorem 3.1]. This completes the proof. □

Remark 3.1. *We note here that several cases of the parameters n, α and β lead to many known results. For instance, as indicated earlier, if we take $\alpha = 1$ in Corollaries 1.2 and 1.3, we have the results of Tuan and Anh [5, Theorems 3.1 and 4.1]. Also for $\beta = 0$, Corollary 1.2 gives a similar result to that of Yamaguchi [6, Theorem 1] while Corollary 1.3 similarly leads to a result of Macgregor [2, Theorem 2].*

Bibliography

[1] K.O. Babalola and T.O. Opoola, Iterated integral transforms of Caratheodory functions and their applications to analytic and univalent functions, *to appear*.

[2] T.H. Macgregor, Function whose derivatives have positive real part, *Trans. Amer. Math. Soc.,* **104** (1962), 532–537.

[3] T.O. Opoola, On a new subclass of univalent function, *Matematika,* **(36) 59**(2) (1994), 195–200.

[4] G.S. Salagean, Subclasses of univalent functions, *Lecture Notes in Math.*, **1013** (1983), 362–372. Springer-Verlag, Berlin, Heidelberg & New York.

[5] P.D. Tuan and V.V. Anh, Radii of starlikeness and convexity of certain classes of analytic functions, *J. Math. Anal. Appl.,* **64** (1978), 146–158.

[6] K. Yamaguchi, On functions satisfying $\text{Re}\,\{f(z)/z\} > 0$, *Proc. Amer. Math. Soc.,* **17** (1966), 588–591.

In: Advances in Inequalities for Series
Editors: S. S. Dragomir and A. Sofo, pp. 23–64
ISBN 978-1-60021-920-7

Chapter 3

Approximations of $f-$Divergence Measures via Taylor Series Expansions: A Survey of Recent Results

N. S. Barnett, P. Cerone, S. S. Dragomir and A. Sofo
School of Computer Science and Mathematics
Victoria University, PO Box 14428 Melbourne
VIC 8001, Australia
E-mail addresses: {neil, pc, sever, sofo}@csm.vu.edu.au

Abstract

Recent approximations of the Csiszár $f-$divergence are surveyed. The approximations involve the use of Taylor's and perturbed Taylor's series. Some applications for Kullback-Leibler divergences are presented.

1991 Mathematics Subject Classification: Primary 94Xxx; Secondary 26D15.

Key words and phrases: Csiszár $f-$divergence, Taylor's formula, Analytic inequalities.

1. Introduction

One of the important issues in many applications of Probability Theory is finding an appropriate measure of *distance* (or *discrimination*) between two probability distributions. A number of divergence measures for this purpose have been proposed and extensively studied by Jeffreys [28], Kullback and Leibler [37], Rényi [49], Havrda and Charvat [25], Kapur [31], Sharma and Mittal [51], Burbea and Rao [7], Rao [48], Lin [39], Csiszár [12], Ali and Silvey [1], Vajda [59], Shioya and Da-te [52] and others (see for example [31] and the references therein).

These measures have been applied in a variety of fields such as: anthropology [48], genetics [45], finance, economics, and political science [50], [54], [55], biology [46], the analysis of contingency tables [26], approximation of probability distributions [9], [32], signal processing [29], [30] and pattern recognition [5], [8]. A number of these measures of distance are specific cases of f-divergence and so further exploration of this concept will have a flow on effect to other measures of distance and to areas in which they are applied.

Assume that a set Γ and the $\sigma-$finite measure μ are given. Consider the set of all probability densities on μ to be $\Omega := \{p | p : \Gamma \to \mathbb{R},\ p(x) \geq 0,\ \int_\Gamma p(x)\, d\mu(x) = 1\}$. The Kullback-Leibler divergence [37] is well known among the information divergences. It is defined as:

$$D_{KL}(p,q) := \int_\Gamma p(x) \log\left[\frac{p(x)}{q(x)}\right] d\mu(x),\ \ p,q \in \Omega, \tag{1.1}$$

where log is to base 2.

In Information Theory and Statistics, various divergences are applied in addition to the Kullback-Leibler divergence. These are the: *variation distance* D_v, *Hellinger distance* D_H [27], χ^2-*divergence* D_{χ^2}, $\alpha-$*divergence* D_α, *Bhattacharyya distance* D_B [6], *Harmonic distance* D_{Ha}, *Jeffreys' distance* D_J [28], *triangular discrimination* D_Δ [56], etc. They are defined as follows:

$$D_v(p,q) := \int_\Gamma |p(x) - q(x)|\, d\mu(x),\ \ p,q \in \Omega; \tag{1.2}$$

$$D_H(p,q) := \int_\Gamma \left|\sqrt{p(x)} - \sqrt{q(x)}\right| d\mu(x),\ \ p,q \in \Omega; \tag{1.3}$$

$$D_{\chi^2}(p,q) := \int_\Gamma p(x)\left[\left(\frac{q(x)}{p(x)}\right)^2 - 1\right] d\mu(x),\ \ p,q \in \Omega; \tag{1.4}$$

$$D_\alpha(p,q) := \frac{4}{1-\alpha^2}\left[1 - \int_\Gamma [p(x)]^{\frac{1-\alpha}{2}} [q(x)]^{\frac{1+\alpha}{2}} d\mu(x)\right],\ \ p,q \in \Omega; \tag{1.5}$$

$$D_B(p,q) := \int_\Gamma \sqrt{p(x)\,q(x)}\, d\mu(x),\ \ p,q \in \Omega; \tag{1.6}$$

$$D_{Ha}(p,q) := \int_\Gamma \frac{2p(x)\,q(x)}{p(x)+q(x)} d\mu(x),\ \ p,q \in \Omega; \tag{1.7}$$

$$D_J(p,q) := \int_\Gamma [p(x) - q(x)] \ln\left[\frac{p(x)}{q(x)}\right] d\mu(x),\ \ p,q \in \Omega; \tag{1.8}$$

$$D_\Delta(p,q) := \int_\Gamma \frac{[p(x) - q(x)]^2}{p(x)+q(x)} d\mu(x),\ \ p,q \in \Omega. \tag{1.9}$$

For other divergence measures, see [31] by Kapur or the book on line [53] by Taneja. For a comprehensive collection of preprints available on line, see the RGMIA web site .

The concept of $f-$divergence is defined as follows [12],

$$D_f(p,q) := \int_\Gamma p(x)\, f\left[\frac{q(x)}{p(x)}\right] d\mu(x),\ \ p,q \in \Omega, \tag{1.10}$$

where f is convex on $(0,\infty)$. It is assumed that $f(u)$ is zero and strictly convex at $u=1$. By appropriately defining this convex function, various divergences are derived. All the above distances $(1.1)-(1.9)$, are particular instances of f−divergence. There are also many others which are not in this class (see for example [31] or [53]). For the basic properties of f−divergence see [11] – [14].

In [40], Lin and Wong (see also [39]) introduced the following divergence,

$$D_{LW}(p,q) := \int_\Gamma p(x)\log\left[\frac{p(x)}{\frac{1}{2}p(x)+\frac{1}{2}q(x)}\right]d\mu(x),\quad p,q\in\Omega. \tag{1.11}$$

This can be represented as follows, using the Kullback-Leibler divergence:

$$D_{LW}(p,q) = D_{KL}\left(p,\frac{1}{2}p+\frac{1}{2}q\right).$$

Lin and Wong have established the following inequalities

$$D_{LW}(p,q) \le \frac{1}{2}D_{KL}(p,q); \tag{1.12}$$

$$D_{LW}(p,q)+D_{LW}(q,p) \le D_v(p,q) \le 2; \tag{1.13}$$

$$D_{LW}(p,q) \le 1. \tag{1.14}$$

In [52], Shioya and Da-te improved $(1.12)-(1.14)$ by showing that,

$$D_{LW}(p,q) \le \frac{1}{2}D_v(p,q) \le 1.$$

For classical and new results in comparing different kinds of divergence measures, see [28] – [52] where further references are given.

The main aim of the present paper is to survey some recent results due to the authors concerning the approximation of f−divergences by the use of Taylor's and perturbed Taylor's series expansions. Applications for the important particular case of Kullback-Leibler divergence are given.

2. Approximation of Kullback-Leibler Distance

2.1. Some Inequalities for the Logarithmic Mapping

The following theorem is well known in the literature as Taylor's formula or Taylor's theorem with the integral remainder.

Theorem 2.1. *Let $I\subset\mathbb{R}$ be a closed interval, $a\in I$ and let n be a positive integer. If $f:I\longrightarrow\mathbb{R}$ is such that $f^{(n)}$ is absolutely continuous on I, then for each $x\in I$*

$$f(x) = T_n(f;a,x)+R_n(f;a,x) \tag{2.1}$$

where $T_n(f;a,x)$ is Taylor's polynomial, i.e.,

$$T_n(f;a,x) := \sum_{k=0}^{n} \frac{(x-a)^k}{k!} f^{(k)}(a).$$

(Note that $f^{(0)} := f$ and $0! := 1$), and the remainder is given by

$$R_n(f;a,x) := \frac{1}{n!} \int_a^x (x-t)^n f^{(n+1)}(t)\,dt.$$

A simple proof of this theorem can be achieved by mathematical induction using the integration by parts formula.

The following corollary concerning the estimation of the remainder is useful when we want to approximate real functions by their Taylor expansions.

Corollary 2.2. *With the above assumptions, we have the estimation*

$$|R_n(f;a,x)| \le \frac{|x-a|^n}{n!} \left| \int_a^x \left| f^{(n+1)}(t) \right| dt \right| \tag{2.2}$$

or

$$|R_n(f;a,x)| \le \frac{1}{n!} \frac{|x-a|^{n+\frac{1}{\beta}}}{(n\beta+1)^{\frac{1}{\beta}}} \left| \int_a^x \left| f^{(n+1)}(t) \right|^{\alpha} dt \right|^{\frac{1}{\alpha}} \tag{2.3}$$

where $\alpha > 1$ and $\frac{1}{\alpha} + \frac{1}{\beta} = 1$, and the estimation:

$$|R_n(f;a,x)| \le \frac{|x-a|^{n+1}}{(n+1)!} \max\left\{ \left| f^{(n+1)}(t) \right|, \ t \in [a,x] \text{ or } [x,a] \right\} \tag{2.4}$$

respectively.

Proof. The inequalities (2.2) and (2.4) are obvious.
Using Hölder's integral inequality, we have that

$$\begin{aligned} \left| \int_a^x (x-t)^n f^{(n+1)}(t)\,dt \right| &\le \left| \int_a^x \left| f^{(n+1)}(t) \right|^{\alpha} dt \right|^{\frac{1}{\alpha}} \left| \int_a^x |x-t|^{n\beta} dt \right|^{\frac{1}{\beta}} \\ &= \left[\frac{|x-a|^{n\beta+1}}{n\beta+1} \right]^{\frac{1}{\beta}} \left| \int_a^x \left| f^{(n+1)}(t) \right|^{\alpha} dt \right|^{\frac{1}{\alpha}} \end{aligned}$$

and the inequality (2.3) is also proved. □

The following result for the logarithmic mapping holds.

Corollary 2.3. *Let $a, b > 0$, then we have the equality:*

$$\ln b - \ln a - \frac{b-a}{a} + \sum_{k=2}^{n} \frac{(-1)^k (b-a)^k}{k a^k} = (-1)^n \int_a^b \frac{(b-t)^n}{t^{n+1}} dt. \tag{2.5}$$

Proof. Consider the mapping $f:(0,\infty)\longrightarrow\mathbb{R}$, $f(x)=\ln x$, then,

$$f^{(n)}(x)=\frac{(-1)^{n-1}(n-1)!}{x^n},\ n\geq 1,\ x>0,$$

$$T_n(f;a,x)=\ln a+\sum_{k=1}^{n}\frac{(-1)^{k-1}(x-a)^k}{ka^k},\ a>0$$

and

$$R_n(f;a,x)=(-1)^n\int_a^x\frac{(x-t)^n}{t^{n+1}}dt.$$

Now, using (2.1), we have the equality,

$$\ln x=\ln a+\sum_{k=1}^{n}\frac{(-1)^{k-1}(x-a)^k}{ka^k}+(-1)^n\int_a^x\frac{(x-t)^n}{t^{n+1}}dt.$$

That is,

$$\ln x-\ln a+\sum_{k=1}^{n}\frac{(-1)^k(x-a)^k}{ka^k}=(-1)^n\int_a^x\frac{(x-t)^n}{t^{n+1}}dt,\quad x,a>0.$$

Choosing $x=b$ in the last equality, we get (2.5). □

The following inequality for logarithms holds.

Corollary 2.4. *For all $a,b>0$, we have the inequalities:*

$$\left|\ln b-\ln a-\frac{b-a}{a}+\sum_{k=2}^{n}\frac{(-1)^k(b-a)^k}{ka^k}\right|$$
$$\leq\begin{cases}\frac{|b-a|^n|b^n-a^n|}{na^nb^n};\\ \frac{|b-a|^{n+\frac{1}{\beta}}}{[(n+1)\alpha-1]^{\frac{1}{\alpha}}(n\beta+1)^{\frac{1}{\beta}}}\left[\frac{\left|b^{(n+1)\alpha-1}-a^{(n+1)\alpha-1}\right|}{b^{(n+1)\alpha-1}a^{(n+1)\alpha-1}}\right]^{\frac{1}{\alpha}},\quad \alpha>1,\ \frac{1}{\alpha}+\frac{1}{\beta}=1;\\ \frac{|b-a|^{n+1}}{n+1}\left[\frac{1}{\min\{a,b\}}\right]^{n+1}.\end{cases}\tag{2.6}$$

Equality holds if and only if $a=b$.

Proof. We use Corollary 2.2 for mapping $f:(0,\infty)\to\mathbb{R}$, $f(x)=\ln x$ for which we have,

$$\int_a^b\left|f^{(n+1)}(t)\right|dt=n!\int_a^b\frac{dt}{t^{n+1}}=n!\left[\frac{t^{-n+1-1}}{-n+1-1}\bigg|_a^b\right]=\frac{n!}{n}\left[\frac{1}{a^n}-\frac{1}{b^n}\right]=\frac{n!}{n}\cdot\frac{b^n-a^n}{a^nb^n}.$$

Also,

$$\int_a^b\left|f^{(n+1)}(t)\right|^\alpha dt=(n!)^\alpha\int_a^b\frac{dt}{t^{\alpha(n+1)}}=\frac{(n!)^\alpha}{(n+1)\alpha-1}\cdot\frac{b^{(n+1)\alpha-1}-a^{(n+1)\alpha-1}}{b^{(n+1)\alpha-1}\cdot a^{(n+1)\alpha-1}}$$

and

$$\max\left\{\left|f^{(n+1)}(t)\right|,\ t\in[a,b] \text{ or } t\in[b,a]\right\}=\max\left\{n!\frac{1}{t^{n+1}},\ t\in[a,b] \text{ or } t\in[b,a]\right\}$$
$$=n!\frac{1}{\min\{a^{n+1},b^{n+1}\}}=n!\left[\frac{1}{\min\{a,b\}}\right]^{n+1}.$$

The equality in (2.6) holds via the representation (2.5), the details are omitted. □

Remark 2.1. *By the concavity property of* $\ln(\cdot)$ *we have*

$$\ln b-\ln a\le\frac{b-a}{a}$$

and so, if we choose $n=1$ *in* (2.6), *we get the following reverse inequality*

$$0\le\frac{b-a}{a}-\ln b+\ln a\le\begin{cases}\frac{(b-a)^2}{ab};\\ \frac{|b-a|^{1+\frac{1}{\beta}}\left|b^{2\alpha-1}-a^{2\alpha-1}\right|^{\frac{1}{\alpha}}}{(2\alpha-1)^{\frac{1}{\alpha}}(\beta+1)^{\frac{1}{\beta}}a^{2-\frac{1}{\alpha}}b^{2-\frac{1}{\alpha}}}, & \alpha>1,\ \frac{1}{\alpha}+\frac{1}{\beta}=1;\\ \frac{(b-a)^2}{2}\cdot\frac{1}{\min^2\{a,b\}}.\end{cases}$$

Equality holds in all inequalities simultaneously if and only if $a=b$.

Remark 2.2. *If we choose* $n=2$ *in* (2.6), *we get,*

$$\left|\ln b-\ln a-\frac{b-a}{a}+\frac{(b-a)^2}{2a^2}\right|\le\begin{cases}\frac{(b-a)^3}{a^2b^2}\cdot\frac{a+b}{2};\\ \frac{|b-a|^{2+\frac{1}{\beta}}\left|b^{3\alpha-1}-a^{3\alpha-1}\right|^{\frac{1}{\alpha}}}{(3\alpha-1)^{\frac{1}{\alpha}}(2\beta+1)^{\frac{1}{\beta}}a^{3-\frac{1}{\alpha}}b^{3-\frac{1}{\alpha}}}, & \alpha>1,\ \frac{1}{\alpha}+\frac{1}{\beta}=1;\\ \frac{(b-a)^3}{3}\cdot\frac{1}{\min^3\{a,b\}}.\end{cases}$$

Equality holds in all inequalities simultaneously if and only if $a=b$.

2.2. Inequalities for Relative Entropy

Let X and Y be two random variables having the probability mass functions $p(x)$, $q(x)$, $x\in\Gamma$, then we have the following representation of relative entropy [20].

Theorem 2.5. *With the above assumptions of* X *and* Y, *we have,*

$$D_{KL}(p,q)=\sum_{k=2}^{n}\int_{\Gamma}\frac{(p(x)-q(x))^k}{kp^{k-1}(x)}d\mu(x)$$
$$+(-1)^{n-1}\int_{\Gamma}p(x)\left(\int_{p(x)}^{q(x)}\frac{(q(x)-t)^n}{t^{n+1}}dt\right)d\mu(x) \qquad (2.7)$$

or

$$D_{KL}(p,q) = \sum_{k=1}^{n} \int_{\Gamma} p(x) \left(\frac{(-1)^{k-1} (p(x) - q(x))^k}{kq^k(x)} \right) d\mu(x)$$
$$+ (-1)^n \int_{\Gamma} p(x) \left(\int_{q(x)}^{p(x)} \frac{(p(x) - t)^n}{t^{n+1}} dt \right) d\mu(x). \qquad (2.8)$$

Proof. Choose $a = p(x)$, $b = q(x)$, $x \in \Gamma$ in (2.5) to get,

$$\ln q(x) - \ln p(x) - \frac{q(x) - p(x)}{p(x)} + \sum_{k=2}^{n} \frac{(-1)^k (q(x) - p(x))^k}{kp^k(x)}$$
$$= (-1)^n \int_{p(x)}^{q(x)} \frac{(q(x) - t)^n}{t^{n+1}} dt. \qquad (2.9)$$

Multiply (2.9) by $p(x)$ and integrate over $x \in \Gamma$ to obtain,

$$-D_{KL}(p,q) - \int_{\Gamma} [q(x) - p(x)] d\mu(x) + \int_{\Gamma} \left(\sum_{k=2}^{n} \frac{(-1)^k (q(x) - p(x))^k}{kp^{k-1}(x)} \right) d\mu(x)$$
$$= (-1)^n \int_{\Gamma} p(x) \left(\int_{p(x)}^{q(x)} \frac{(q(x) - t)^n}{t^{n+1}} dt \right) d\mu(x). \qquad (2.10)$$

However,

$$\int_{\Gamma} [q(x) - p(x)] d\mu(x) = 0,$$

therefore, by (2.10) we get (2.7).

To prove the second identity (2.8), choose $b = p(x)$, $a = q(x)$, $x \in \Gamma$ in (2.5) to get,

$$\ln p(x) - \ln q(x) - \frac{p(x) - q(x)}{q(x)} + \sum_{k=2}^{n} \frac{(-1)^k (p(x) - q(x))^k}{kq^k(x)}$$
$$= (-1)^n \int_{q(x)}^{p(x)} \frac{(p(x) - t)^n}{t^{n+1}} dt. \qquad (2.11)$$

Multiply (2.11) by $p(x)$ and sum over $x \in \Gamma$ to get,

$$D_{KL}(p,q) = \int_{\Gamma} p(x) \left(\sum_{k=1}^{n} \frac{(-1)^{k-1} (p(x) - q(x))^k}{kq^k(x)} \right) d\mu(x)$$
$$+ (-1)^n \int_{\Gamma} p(x) \left(\int_{q(x)}^{p(x)} \frac{(p(x) - t)^n}{t^{n+1}} dt \right) d\mu(x)$$

from which we deduce (2.8). □

Using Corollary 2.4, we can give the following result containing an approximation of the relative entropy [20].

Theorem 2.6. *With the above assumption over X and Y, we have,*

$$\left|D_{KL}(p,q)-\sum_{k=2}^{n}\int_{\Gamma}\left(\frac{(p(x)-q(x))^{k}}{kp^{k-1}(x)}\right)d\mu(x)\right|$$

$$\leq M:=\begin{cases}\frac{1}{n}\int_{\Gamma}\frac{|q(x)-p(x)|^{n}|q^{n}(x)-p^{n}(x)|}{p^{n-1}(x)q^{n}(x)}d\mu(x);\\ \frac{1}{[(n+1)\alpha-1]^{\frac{1}{\alpha}}(n\beta+1)^{\frac{1}{\beta}}}\int_{\Gamma}p(x)|q(x)-p(x)|^{n+\frac{1}{\beta}}\\ \quad\times\left[\frac{|q^{(n+1)\alpha-1}(x)-p^{(n+1)\alpha-1}(x)|}{q^{(n+1)\alpha-1}(x)p^{(n+1)\alpha-1}(x)}\right]^{\frac{1}{\alpha}}d\mu(x);\\ \frac{1}{n+1}\int_{\Gamma}p(x)|q(x)-p(x)|^{n+1}\left[\frac{1}{\min\{p(x),q(x)\}}\right]^{n+1}d\mu(x);\end{cases}$$

and

$$\left|D_{KL}(p,q)-\int_{\Gamma}p(x)\left(\sum_{k=1}^{n}\frac{(-1)^{k-1}(p(x)-q(x))^{k}}{kq^{k}(x)}\right)d\mu(x)\right|\leq M.$$

Proof. Proof of the first inequality is obvious by Corollary 2.4, choosing $a=p(x), b=q(x)$, multiplying by $q(x)$ and integrating over $x\in\Gamma$. Proof for second inequality is obvious by Corollary 2.4, choosing $b=p(x)$, $a=q(x)$, multiplying by $q(x)$ and integrating over $x\in\Gamma$. □

Corollary 2.7. *Under the assumptions of Theorem 2.6 for $n=1$, we have,*

$$D_{KL}(p,q)\leq M_1 \tag{2.12}$$

where

$$M_1:=\begin{cases}\int_{\Gamma}\frac{(q(x)-p(x))^{2}}{q(x)}d\mu(x)=D_{\chi^2}(q,p);\\ \frac{1}{(2\alpha-1)^{\frac{1}{\alpha}}(\beta+1)^{\frac{1}{\beta}}}\int_{\Gamma}p(x)|q(x)-p(x)|^{1+\frac{1}{\beta}}\times\frac{|q^{2\alpha-1}(x)-p^{2\alpha-1}(x)|^{\frac{1}{\alpha}}}{q^{2-\frac{1}{\alpha}}(x)p^{2-\frac{1}{\alpha}}(x)}d\mu(x);\\ \frac{1}{2}\int_{\Gamma}p(x)(q(x)-p(x))^{2}\times\frac{1}{\min^{2}\{p(x),q(x)\}}d\mu(x);\end{cases}$$

and

$$0\leq D_{\chi^2}(q,p)-D_{KL}(p,q)\leq M_1, \tag{2.13}$$

where $\alpha>1$ and $\frac{1}{\alpha}+\frac{1}{\beta}=1$.

Remark 2.3. *The first inequality in (2.12) is equivalent to (see also [16]):*

$$D_{KL}(p,q)\leq D_{\chi^2}(q,p)-1.$$

We introduce the notation $M_{\alpha,\beta}$ *for the second term in* M_1 *and apply Hölder's inequality. We can then write,*

$$
\begin{aligned}
M_{\alpha,\beta} &:= \int_\Gamma p(x)\left|q(x)-p(x)\right|^{\frac{\beta+1}{\beta}} \times \frac{\left|q^{2\alpha-1}(x)-p^{2\alpha-1}(x)\right|^{\frac{1}{\alpha}}}{q^{2-\frac{1}{\alpha}}(x)\,p^{2-\frac{1}{\alpha}}(x)} d\mu(x) \\
&\le \left(\int_\Gamma p(x)\left|q(x)-p(x)\right|^{\beta+1} d\mu(x)\right)^{\frac{1}{\beta}} \left(\int_\Gamma \frac{\left|q^{2\alpha-1}(x)-p^{2\alpha-1}(x)\right|}{q^{2\alpha-1}(x)\,p^{2\alpha-2}(x)} d\mu(x)\right)^{\frac{1}{\alpha}} \\
&:= \tilde{M}_{\alpha,\beta}
\end{aligned}
$$

and from the second inequality of (2.12) we obtain,

$$
D_{KL}(p,q) \le \frac{1}{(2\alpha-1)^{\frac{1}{\alpha}}(\beta+1)^{\frac{1}{\beta}}} \tilde{M}_{\alpha,\beta}.
$$

For $\alpha=\beta=2$*, we get the particular inequality,*

$$
D_{KL}(p,q) \le \frac{1}{3}\left(\int_\Gamma p(x)\left|q(x)-p(x)\right|^3 d\mu(x)\right)^{\frac{1}{2}} \left(\int_\Gamma \frac{\left|q^3(x)-p^3(x)\right|}{q^3(x)\,p^2(x)} d\mu(x)\right)^{\frac{1}{2}}.
$$

If we assume that

$$
\inf_{x\in\Gamma}\left[\min\{p(x),q(x)\}\right] = \delta > 0 \tag{2.14}
$$

then from the third inequality of (2.12) we have,

$$
D_{KL}(p,q) \le \frac{1}{2\delta^2}\int_\Gamma p(x)\left(q(x)-p(x)\right)^2 d\mu(x).
$$

Remark 2.4. *Since*

$$
\int_\Gamma \frac{p(x)}{q(x)}\left(p(x)-q(x)\right) d\mu(x) = D_{\chi^2}(q,p),
$$

then the first inequality in (2.13) is obvious.
Using Hölder's inequality, from the second inequality in (2.13) we get,

$$
D_{\chi^2}(q,p) - D_{KL}(p,q) \le \frac{1}{(2\alpha-1)^{\frac{1}{\alpha}}(\beta+1)^{\frac{1}{\beta}}} \tilde{M}_{\alpha,\beta}.
$$

If, as above, we assume that (2.14) holds, then from the third inequality in (2.13) we have,

$$
D_{\chi^2}(q,p) - D_{KL}(p,q) \le \frac{1}{2\delta^2}\int_\Gamma p(x)\left(q(x)-p(x)\right)^2 d\mu(x).
$$

2.3. Inequalities for the Entropy Mapping

Let X be a random variable having the probability mass function $p(x), x \in \Gamma$. Consider the *entropy mapping,*

$$H(X) = \int_\Gamma p(x) \ln\left(\frac{1}{p(x)}\right) d\mu(x).$$

We have the following representation of $H(X)$ [20].

Theorem 2.8. *With the above assumption for X and if $\mu(\Gamma) < \infty$, we have,*

$$H(X) = \ln[\mu(\Gamma)] - \sum_{k=2}^{n} \int_\Gamma \left[\frac{(\mu(\Gamma)p(x) - 1)^k}{k\mu(\Gamma)^k p^{k-1}(x)}\right] d\mu(x) + \frac{(-1)^n}{[\mu(\Gamma)]^n} \int_\Gamma p(x) \left(\int_{p(x)}^{\frac{1}{\mu(\Gamma)}} \frac{(1-\mu(\Gamma)t)^n}{t^{n+1}} dt\right) d\mu(x) \quad (2.15)$$

or

$$H(X) = \ln[\mu(\Gamma)] + \sum_{k=1}^{n} \int_\Gamma p(x) \left(\frac{(-1)^k (\mu(\Gamma)p(x) - 1)^k}{k}\right) d\mu(x) + (-1)^{n+1} \int_\Gamma p(x) \left(\int_{\frac{1}{\mu(\Gamma)}}^{p(x)} \frac{(p(x)-t)^n}{t^{n+1}} dt\right) d\mu(x). \quad (2.16)$$

Proof. Put $q = u$ in (2.7), where u is the uniform distribution on Γ, i.e., $u(x) = \frac{1}{\mu(\Gamma)}$, then,

$$\begin{aligned} &\ln[\mu(\Gamma)] - H(X) \\ &= \sum_{k=2}^{n} \int_\Gamma \frac{\left(p(x) - \frac{1}{\mu(\Gamma)}\right)^k}{kp^{k-1}(x)} d\mu(x) + (-1)^{n-1} \int_\Gamma p(x) \left(\int_{p(x)}^{\frac{1}{\mu(\Gamma)}} \frac{\left(\frac{1}{\mu(\Gamma)} - t\right)^n}{t^{n+1}} dt\right) d\mu(x) \\ &= \sum_{k=2}^{n} \int_\Gamma \frac{(\mu(\Gamma)p(x) - 1)^k}{k[\mu(\Gamma)]^k p^{k-1}(x)} + \frac{(-1)^{n-1}}{[\mu(\Gamma)]^n} \int_\Gamma p(x) \left(\int_{p(x)}^{\frac{1}{\mu(\Gamma)}} \frac{(1-\mu(\Gamma)t)^n}{t^{n+1}} dt\right) d\mu(x) \end{aligned}$$

from which results (2.15).
Put $q = u$ in (2.8) to get,

$$\begin{aligned} \ln[\mu(\Gamma)] - H(X) &= \int_\Gamma p(x) \sum_{k=1}^{n} \frac{(-1)^{k-1}\left(p(x) - \frac{1}{\mu(\Gamma)}\right)^k}{k\frac{1}{[\mu(\Gamma)]^k}} d\mu(x) \\ &\quad + (-1)^n \int_\Gamma p(x) \left(\int_{\frac{1}{\mu(\Gamma)}}^{p(x)} \frac{(p(x)-t)^n}{t^{n+1}} dt\right) d\mu(x) \\ &= \sum_{k=1}^{n} \int_\Gamma p(x) \frac{(-1)^{k-1}(\mu(\Gamma)p(x) - 1)^k}{k} d\mu(x) \\ &\quad + (-1)^n \int_\Gamma p(x) \left(\int_{\frac{1}{\mu(\Gamma)}}^{p(x)} \frac{(p(x)-t)^n}{t^{n+1}} dt\right) d\mu(x), \end{aligned}$$

from which results (2.16). □

Using Theorem 2.6, we can state the following result concerning the approximation of the entropy mapping [20].

Theorem 2.9. *With the above assumption for X and with $\mu(\Gamma)<\infty$, we have,*

$$\left|H(X)-\ln[\mu(\Gamma)]+\sum_{k=2}^{n}\int_{\Gamma}\frac{(\mu(\Gamma)p(x)-1)^k}{k\mu(\Gamma)^k p^{k-1}(x)}d\mu(x)\right|$$

$$\leq\mu:=\begin{cases}\frac{1}{n[\mu(\Gamma)]^n}\int_\Gamma\frac{|1-\mu(\Gamma)p(x)|^n|1-[\mu(\Gamma)]^n p^n(x)|}{p^{n-1}(x)}d\mu(x);\\ \frac{1}{[(n+1)\alpha-1]^{\frac{1}{\alpha}}(n\beta+1)^{\frac{1}{\beta}}}\int_\Gamma p(x)\left|\frac{1}{\mu(\Gamma)}-p(x)\right|^{n+\frac{1}{\beta}}\left[\frac{\left|1-[\mu(\Gamma)]^{(n+1)\alpha-1}p^{(n+1)\alpha-1}(x)\right|}{p^{(n+1)\alpha-1}(x)}\right]^{\frac{1}{\alpha}}d\mu(x);\\ \frac{1}{n+1}\int_\Gamma p(x)\left|\frac{1}{\mu(\Gamma)}-p(x)\right|^{n+1}\times\left[\frac{1}{\min\left\{p(x),\frac{1}{\mu(\Gamma)}\right\}}\right]^{n+1}d\mu(x);\end{cases}$$

and

$$\left|H(X)-\ln[\mu(\Gamma)]-\int_\Gamma p(x)\sum_{k=1}^{n}\frac{(-1)^k(\mu(\Gamma)p(x)-1)^k}{k}d\mu(x)\right|\leq\mu.$$

2.4. Inequalities for Mutual Information

Let X and Y be random variables having the probability mass functions $p(x), q(y), x\in\Gamma$, $y\in\Gamma$. Consider the *mutual information* [10],

$$I(X,Y)=\int_\Gamma\int_\Gamma p(x,y)\ln\frac{p(x,y)}{p(x)q(y)}d\mu(x)d\mu(y).$$

We have the following representation for $I(X,Y)$ [20].

Theorem 2.10. *With the above assumption for X and Y,*

$$I(X,Y)=\sum_{k=2}^{n}\int_\Gamma\int_\Gamma\frac{(p(x,y)-p(x)q(y))^k}{kp^{k-1}(x,y)}d\mu(x)d\mu(y)$$
$$+(-1)^{n-1}\int_\Gamma\int_\Gamma p(x,y)\left(\int_{p(x,y)}^{p(x)q(y)}\frac{(p(x)q(y)-t)^n}{t^{n+1}}dt\right)d\mu(x)d\mu(y)$$

or

$$I(X,Y)=\sum_{k=1}^{n}\int_\Gamma\int_\Gamma p(x,y)\left(\frac{(-1)^{k-1}(p(x,y)-p(x)q(y))^k}{kp^k(x)q^k(y)}\right)d\mu(x)d\mu(y)$$
$$+(-1)^n\int_\Gamma\int_\Gamma p(x,y)\left(\int_{p(x)q(y)}^{p(x,y)}\frac{(p(x,y)-t)^n}{t^{n+1}}dt\right)d\mu(x)d\mu(y).$$

The proof follows by Theorem 2.5, taking into account that

$$I(X,Y)=D(p(x,y)\|p(x)q(y)).$$

Finally, using Theorem 2.6, we can state the following estimation of the *mutual information.*

Theorem 2.11. *With the above assumption over X and Y, we have:*

$$\left| I(X,Y) - \sum_{k=2}^{n} \int_{\Gamma}\int_{\Gamma} \frac{(p(x,y)-p(x)q(y))^{k}}{kp^{k-1}(x,y)} d\mu(x)\, d\mu(y) \right|$$

$$\leq \tilde{M} := \begin{cases} \frac{1}{n}\int_{\Gamma}\int_{\Gamma} \frac{|p(x)q(y)-p(x,y)|^{n}}{p^{n-1}(x,y)p^{n-1}(x)q^{n-1}(y)} \times |p^{n}(x)q^{n}(y) - p^{n}(x,y)|\, d\mu(x)\, d\mu(y); \\[2ex] \frac{1}{[(n+1)\alpha-1]^{\frac{1}{\alpha}}(n\beta+1)^{\frac{1}{\beta}}} \int_{\Gamma}\int_{\Gamma} p(x,y) \times |p(x)q(y)-p(x,y)|^{n+\frac{1}{\beta}} \\ \qquad \times \left[\frac{|p^{(n+1)\alpha-1}(x)q^{(n+1)\alpha-1}(y) - p^{(n+1)\alpha-1}(x,y)|}{p^{(n+1)\alpha-1}(x)q^{(n+1)\alpha-1}(y)p^{(n+1)\alpha-1}(x,y)}\right]^{\frac{1}{\alpha}} d\mu(x)\, d\mu(y); \\[2ex] \frac{1}{n+1}\int_{\Gamma}\int_{\Gamma} p(x,y) \times |p(x)q(y)-p(x,y)|^{n+1} \\ \qquad \times \left[\frac{1}{\min\{p(x,y),p(x)q(y)\}}\right]^{n+1} d\mu(x)\, d\mu(y); \end{cases}$$

and

$$\left| I(X,Y) - \sum_{k=1}^{n} \int_{\Gamma}\int_{\Gamma} p(x,y) \frac{(-1)^{k-1}(p(x,y)-p(x)q(y))^{k}}{kp^{k}(x)q^{k}(y)} d\mu(x)\, d\mu(y) \right| \leq \tilde{M}.$$

3. Approximation of $f-$Divergences

3.1. General Results

We start with the following result [3].

Theorem 3.1. *Let $f:(0,\infty)\to\mathbb{R}$ be $n-$time differentiable and such that $f^{(n)}$ is absolutely continuous on $[r,R]$, where $0<r\leq 1\leq R<\infty$. Assume that the probability distributions p,q satisfy the condition,*

$$r \leq \frac{p(x)}{q(x)} \leq R \quad a.e.\ on\ \Gamma. \tag{3.1}$$

We then have,

$$\left| I_f(p,q) - f(1) - \sum_{k=1}^{n} \frac{f^{(k)}(1)}{k!} D_{\chi^k}(p,q) \right| \tag{3.2}$$

$$\leq \begin{cases} \frac{1}{(n+1)!} \left\| f^{(n+1)} \right\|_{\infty} D_{|\chi|^{n+1}}(p,q) & \text{if } f^{(n+1)} \in L_{\infty}[r,R]; \\ \frac{1}{n!(n\beta+1)^{\frac{1}{\beta}}} \left\| f^{(n+1)} \right\|_{\alpha} D_{|\chi|^{n+\frac{1}{\beta}}}(p,q) & \text{if } f^{(n+1)} \in L_{\alpha}[r,R], \\ & \alpha > 1,\ \frac{1}{\alpha}+\frac{1}{\beta}=1; \\ \frac{1}{n!} \left\| f^{(n+1)} \right\|_{1} D_{|\chi|^{n}}(p,q); \end{cases}$$

$$\leq \begin{cases} \frac{1}{(n+1)!} \left\| f^{(n+1)} \right\|_{\infty} (R-r)^{n+1} & \text{if } f^{(n+1)} \in L_{\infty}[r,R]; \\ \frac{1}{n!(n\beta+1)^{\frac{1}{\beta}}} \left\| f^{(n+1)} \right\|_{\alpha} (R-r)^{n+\frac{1}{\beta}} & \text{if } f^{(n+1)} \in L_{\alpha}[r,R], \\ & \alpha > 1,\ \frac{1}{\alpha}+\frac{1}{\beta}=1; \\ \frac{1}{n!} \left\| f^{(n+1)} \right\|_{1} (R-r)^{n}; \end{cases}$$

where

$$D_{\chi^k}(p,q) := \int_{\Gamma} \frac{(p(x)-q(x))^k}{q^{k-1}(x)} d\mu(x),$$

$$D_{|\chi|^s}(p,q) := \int_{\Gamma} \frac{|p(x)-q(x)|^s}{q^{s-1}(x)} d\mu(x) \quad (k \in \mathbb{N},\ s \geq 0)$$

and $\|\cdot\|_{\alpha}$ *are the usual Lebesgue norms, i.e.,*

$$\left\| f^{(n+1)} \right\|_{\alpha} := \left(\int_r^R \left| f^{(n+1)} \right|^{\alpha} dt \right)^{\frac{1}{\alpha}},\ \alpha \geq 1,$$

$$\left\| f^{(n+1)} \right\|_{\infty} := ess \sup_{t \in [r,R]} \left| f^{(n+1)}(t) \right|.$$

Proof. We start with a Taylor expansion with integral remainder:

$$f(z) = f(a) + \sum_{k=1}^{n} \frac{(z-a)^k}{k!} f^{(k)}(a) + \frac{1}{n!} \int_a^z (z-t)^n f^{(n+1)}(t)\,dt \tag{3.3}$$

for all $z, a \in (0,\infty)$.

Using the properties of the modulus, we have,

$$\left| f(z) - f(a) - \sum_{k=1}^{n} \frac{(z-a)^k}{k!} f^{(k)}(a) \right| \leq \frac{1}{n!} \left| \int_a^z |z-t|^n \left| f^{(n+1)}(t) \right| dt \right| \tag{3.4}$$

$$:= M\left(f^{(n+1)}; a, z \right).$$

Now, assume that $a, z \in [r, R]$, then, obviously,

$$M\left(f^{(n+1)}; a, z\right) \leq \sup_{t\in[r,R]} \left|f^{(n+1)}(t)\right| \frac{1}{n!} \left|\int_a^z |z-t|^n \, dt\right| = \frac{1}{(n+1)!} \left\|f^{(n+1)}\right\|_\infty |z-a|^{n+1}, \tag{3.5}$$

for all $a, z \in [r, R]$.

Also, by use of Hölder's integral inequality, we have:

$$\begin{aligned} M\left(f^{(n+1)}; a, z\right) &\leq \frac{1}{n!} \left|\int_a^z |x-t|^{n\beta} \, dt\right|^{\frac{1}{\beta}} \left[\int_a^z \left|f^{(n+1)}(t)\right|^\alpha dt\right]^{\frac{1}{\alpha}} \\ &\leq \frac{1}{n!} \left\|f^{(n+1)}\right\|_\alpha \left[\frac{|z-a|^{n\beta+1}}{n\beta+1}\right]^{\frac{1}{\beta}} \\ &= \frac{1}{n!\,(n\beta+1)^{\frac{1}{\beta}}} \left\|f^{(n+1)}\right\|_\alpha |z-a|^{n+\frac{1}{\beta}}, \quad \alpha > 1, \frac{1}{\alpha} + \frac{1}{\beta} = 1 \end{aligned} \tag{3.6}$$

and, obviously,

$$M\left(f^{(n+1)}; a, z\right) \leq \frac{1}{n!} |z-a|^n \left|\int_a^z \left|f^{(n+1)}(t)\right| dt\right| \leq \frac{1}{n!} |z-a|^n \left\|f^{(n+1)}\right\|_1 \tag{3.7}$$

for all $z, a \in [r, R]$.

Consequently, by (3.4)-(3.7), we may write (see also [18] for a similar inequality),

$$\left|f(z) - f(a) - \sum_{k=1}^{n} \frac{(z-a)^k}{k!} f^{(k)}(a)\right| \leq \begin{cases} \frac{1}{(n+1)!} \left\|f^{(n+1)}\right\|_\infty |z-a|^{n+1} & \text{if } f^{(n+1)} \in L_\infty[r,R]; \\ \frac{1}{n!(n\beta+1)^{\frac{1}{\beta}}} \left\|f^{(n+1)}\right\|_\alpha |z-a|^{n+\frac{1}{\beta}} & \text{if } f^{(n+1)} \in L_\alpha[r,R], \\ & \alpha > 1, \ \frac{1}{\alpha} + \frac{1}{\beta} = 1; \\ \frac{1}{n!} \left\|f^{(n+1)}\right\|_1 |z-a|^n, \end{cases} \tag{3.8}$$

for all $z, a \in [r, R]$.

If in (3.8) we choose $z = \frac{p(x)}{q(x)}$, $a = 1$, then we obtain,

$$\left|f\left(\frac{p(x)}{q(x)}\right) - f(1) - \sum_{k=1}^{n} \frac{\left(\frac{p(x)}{q(x)} - 1\right)^k}{k!} f^{(k)}(1)\right| \tag{3.9}$$

$$\leq \begin{cases} \frac{1}{(n+1)!} \left\|f^{(n+1)}\right\|_\infty \left|\frac{p(x)}{q(x)} - 1\right|^{n+1} \\ \frac{1}{n!(n\beta+1)^{\frac{1}{\beta}}} \left\|f^{(n+1)}\right\|_\alpha \left|\frac{p(x)}{q(x)} - 1\right|^{n+\frac{1}{\beta}} \\ \frac{1}{n!} \left\|f^{(n+1)}\right\|_1 \left|\frac{p(x)}{q(x)} - 1\right|^n, \end{cases} \leq \begin{cases} \frac{1}{(n+1)!} \left\|f^{(n+1)}\right\|_\infty (R-r)^{n+1} \\ \frac{1}{n!(n\beta+1)^{\frac{1}{\beta}}} \left\|f^{(n+1)}\right\|_\alpha (R-r)^{n+\frac{1}{\beta}} \\ \frac{1}{n!} \left\|f^{(n+1)}\right\|_1 (R-r)^n, \end{cases}$$

for a.e. $x \in \Gamma$.

If we multiply (3.9) by $q(x) \geq 0$ and integrate on Γ, and then use the generalised triangle inequality,

$$\left| I_f(p,q) - f(1) - \sum_{k=1}^{n} \frac{f^{(k)}(1)}{k!} \cdot \int_\Gamma \frac{(p(x)-q(x))^k}{q^{k-1}(x)} d\mu(x) \right|$$

$$\leq \begin{cases} \frac{1}{(n+1)!} \left\| f^{(n+1)} \right\|_\infty \int_\Gamma \frac{|p(x)-q(x)|^{n+1}}{q^n(x)} d\mu(x) \\ \frac{1}{n!(n\beta+1)^{\frac{1}{\beta}}} \left\| f^{(n+1)} \right\|_\alpha \int_\Gamma \frac{|p(x)-q(x)|^{n+\frac{1}{\beta}}}{q^{n+\frac{1}{\beta}-1}(x)} d\mu(x) \\ \frac{1}{n!} \left\| f^{(n+1)} \right\|_1 \int_\Gamma \frac{|p(x)-q(x)|^n}{q^{n-1}(x)} d\mu(x), \end{cases} \leq \begin{cases} \frac{1}{(n+1)!} \left\| f^{(n+1)} \right\|_\infty (R-r)^{n+1} \\ \frac{1}{n!(n\beta+1)^{\frac{1}{\beta}}} \left\| f^{(n+1)} \right\|_\alpha (R-r)^{n+\frac{1}{\beta}} \\ \frac{1}{n!} \left\| f^{(n+1)} \right\|_1 (R-r)^n, \end{cases}$$

and the inequality (3.1) is proved. □

The following theorem also holds [3].

Theorem 3.2. *Let f be as in Theorem 3.1. If $p^{(j)}, q^{(j)}$ $(j = 1,2)$ are probability distributions such that,*

$$r \leq \frac{p^{(j)}(x)}{q^{(j)}(x)} \leq R \text{ a.e. on } \Gamma \text{ and } j = 1,2, \tag{3.10}$$

then

$$r \leq \frac{\lambda p^{(1)}(x) + (1-\lambda) p^{(2)}(x)}{\lambda q^{(1)}(x) + (1-\lambda) q^{(2)}(x)} \leq R \text{ a.e. on } \Gamma \text{ and } \lambda \in [0,1] \tag{3.11}$$

and we have the inequality, for $f^{(n+1)} \in L_\alpha[r,R]$,

$$\begin{aligned} &\Big| I_f\left(\lambda p^{(1)} + (1-\lambda) p^{(2)}, \lambda q^{(1)} + (1-\lambda) q^{(2)}\right) \\ &\quad - \lambda I_f\left(p^{(1)}, q^{(1)}\right) - (1-\lambda) I_f\left(p^{(2)}, q^{(2)}\right) \\ &\quad - \lambda \sum_{k=1}^{n} \frac{1}{k!} \int_\Gamma \frac{(1-\lambda)^k}{\left[q^{(1)}(x)\right]^{k-1}} (-1)^k \eta^k(x) f^{(k)}\left(\frac{p^{(1)}(x)}{q^{(1)}(x)}\right) d\mu(x) \\ &\quad - (1-\lambda) \sum_{k=1}^{n} \frac{1}{k!} \int_\Gamma \frac{\lambda^k}{\left[q^{(2)}(x)\right]^{k-1}} \eta^k(x) f^{(k)}\left(\frac{p^{(2)}(x)}{q^{(2)}(x)}\right) d\mu(x) \Big| \end{aligned} \tag{3.12}$$

$$\leq \lambda(1-\lambda) \begin{cases} \frac{1}{(n+1)!} \left\| f^{(n+1)} \right\|_\infty \int_\Gamma |\eta(x)|^{n+1} \left[\frac{(1-\lambda)^n}{\left[q^{(1)}(x)\right]^n} + \frac{\lambda^n}{\left[q^{(2)}(x)\right]^n} \right] d\mu(x); \\ \frac{1}{n!(n\beta+1)^{\frac{1}{\beta}}} \left\| f^{(n+1)} \right\|_\alpha \int_\Gamma |\eta(x)|^{n+\frac{1}{\beta}} \left[\frac{(1-\lambda)^{n+\frac{1}{\beta}-1}}{\left[q^{(1)}(x)\right]^{n+\frac{1}{\beta}-1}} + \frac{\lambda^{n+\frac{1}{\beta}-1}}{\left[q^{(2)}(x)\right]^{n+\frac{1}{\beta}-1}} \right] d\mu(x); \\ \frac{1}{n!} \left\| f^{(n+1)} \right\|_1 \int_\Gamma |\eta(x)|^n \left[\frac{(1-\lambda)^{n-1}}{\left[q^{(1)}(x)\right]^{n-1}} + \frac{\lambda^{n-1}}{\left[q^{(2)}(x)\right]^{n-1}} \right] d\mu(x), \end{cases}$$

where

$$\eta(x) = \eta\left(\lambda, p^{(1)}, p^{(2)}, q^{(1)}, q^{(2)}\right)(x) = \frac{\det \begin{bmatrix} p^{(1)}(x) & p^{(2)}(x) \\ q^{(1)}(x) & q^{(2)}(x) \end{bmatrix}}{\lambda q^{(1)}(x) + (1-\lambda) q^{(2)}(x)},$$

for all $\lambda \in [0,1]$ *and* $x \in \Gamma$.

Proof. We use the inequality (3.8) to write,

$$\begin{aligned}&\left| f\left(\frac{\lambda p^{(1)}(x)+(1-\lambda)p^{(2)}(x)}{\lambda q^{(1)}(x)+(1-\lambda)q^{(2)}(x)}\right) - f\left(\frac{p^{(1)}(x)}{q^{(1)}(x)}\right)\right.\\&\qquad\left. - \sum_{k=1}^{n}\frac{1}{k!}\left(\frac{\lambda p^{(1)}(x)+(1-\lambda)p^{(2)}(x)}{\lambda q^{(1)}(x)+(1-\lambda)q^{(2)}(x)} - \frac{p^{(1)}(x)}{q^{(1)}(x)}\right)^{k} f^{(k)}\left(\frac{p^{(1)}(x)}{q^{(1)}(x)}\right)\right|\\&\le\begin{cases}\frac{1}{(n+1)!}\left\|f^{(n+1)}\right\|_{\infty}\left|\frac{\lambda p^{(1)}(x)+(1-\lambda)p^{(2)}(x)}{\lambda q^{(1)}(x)+(1-\lambda)q^{(2)}(x)} - \frac{p^{(1)}(x)}{q^{(1)}(x)}\right|^{n+1}\\ \frac{1}{n!(n\beta+1)^{\frac{1}{\beta}}}\left\|f^{(n+1)}\right\|_{\alpha}\left|\frac{\lambda p^{(1)}(x)+(1-\lambda)p^{(2)}(x)}{\lambda q^{(1)}(x)+(1-\lambda)q^{(2)}(x)} - \frac{p^{(1)}(x)}{q^{(1)}(x)}\right|^{n+\frac{1}{\beta}}\\ \frac{1}{n!}\left\|f^{(n+1)}\right\|_{1}\left|\frac{\lambda p^{(1)}(x)+(1-\lambda)p^{(2)}(x)}{\lambda q^{(1)}(x)+(1-\lambda)q^{(2)}(x)} - \frac{p^{(1)}(x)}{q^{(1)}(x)}\right|^{n},\end{cases}\end{aligned} \tag{3.13}$$

and

$$\begin{aligned}&\left| f\left(\frac{\lambda p^{(1)}(x)+(1-\lambda)p^{(2)}(x)}{\lambda q^{(1)}(x)+(1-\lambda)q^{(2)}(x)}\right) - f\left(\frac{p^{(2)}(x)}{q^{(2)}(x)}\right)\right.\\&\qquad\left. - \sum_{k=1}^{n}\frac{1}{k!}\left(\frac{\lambda p^{(1)}(x)+(1-\lambda)p^{(2)}(x)}{\lambda q^{(1)}(x)+(1-\lambda)q^{(2)}(x)} - \frac{p^{(2)}(x)}{q^{(2)}(x)}\right)^{k} f^{(k)}\left(\frac{p^{(2)}(x)}{q^{(2)}(x)}\right)\right|\\&\le\begin{cases}\frac{1}{(n+1)!}\left\|f^{(n+1)}\right\|_{\infty}\left|\frac{\lambda p^{(1)}(x)+(1-\lambda)p^{(2)}(x)}{\lambda q^{(1)}(x)+(1-\lambda)q^{(2)}(x)} - \frac{p^{(2)}(x)}{q^{(2)}(x)}\right|^{n+1}\\ \frac{1}{n!(n\beta+1)^{\frac{1}{\beta}}}\left\|f^{(n+1)}\right\|_{\alpha}\left|\frac{\lambda p^{(1)}(x)+(1-\lambda)p^{(2)}(x)}{\lambda q^{(1)}(x)+(1-\lambda)q^{(2)}(x)} - \frac{p^{(2)}(x)}{q^{(2)}(x)}\right|^{n+\frac{1}{\beta}}\\ \frac{1}{n!}\left\|f^{(n+1)}\right\|_{1}\left|\frac{\lambda p^{(1)}(x)+(1-\lambda)p^{(2)}(x)}{\lambda q^{(1)}(x)+(1-\lambda)q^{(2)}(x)} - \frac{p^{(2)}(x)}{q^{(2)}(x)}\right|^{n}.\end{cases}\end{aligned} \tag{3.14}$$

If we multiply (3.13) by $\lambda q^{(1)}(x)$ and (3.14) by $(1-\lambda)q^{(2)}(x)$, add them and apply the triangle inequality to the resulting expression, we obtain,

$$\begin{aligned}&\left|\left(\lambda q^{(1)}(x)+(1-\lambda)q^{(2)}(x)\right) f\left(\frac{\lambda p^{(1)}(x)+(1-\lambda)p^{(2)}(x)}{\lambda q^{(1)}(x)+(1-\lambda)q^{(2)}(x)}\right)\right.\\&\qquad -\lambda q^{(1)}(x) f\left(\frac{p^{(1)}(x)}{q^{(1)}(x)}\right) - (1-\lambda)q^{(2)}(x) f\left(\frac{p^{(2)}(x)}{q^{(2)}(x)}\right)\\&\quad -\lambda q^{(1)}(x)\sum_{k=1}^{n}\frac{1}{k!}\left(\frac{\lambda p^{(1)}(x)+(1-\lambda)p^{(2)}(x)}{\lambda q^{(1)}(x)+(1-\lambda)q^{(2)}(x)} - \frac{p^{(1)}(x)}{q^{(1)}(x)}\right)^{k} f^{(k)}\left(\frac{p^{(1)}(x)}{q^{(1)}(x)}\right)\\&\quad\left. -(1-\lambda)\sum_{k=1}^{n}\frac{1}{k!}\left(\frac{\lambda p^{(1)}(x)+(1-\lambda)p^{(2)}(x)}{\lambda q^{(1)}(x)+(1-\lambda)q^{(2)}(x)} - \frac{p^{(2)}(x)}{q^{(2)}(x)}\right)^{k} f^{(k)}\left(\frac{p^{(2)}(x)}{q^{(2)}(x)}\right)\right|\end{aligned} \tag{3.15}$$

$$\leq \begin{cases} \frac{1}{(n+1)!}\left\|f^{(n+1)}\right\|_{\infty}\left[\lambda q^{(1)}(x)\left|\frac{\lambda p^{(1)}(x)+(1-\lambda)p^{(2)}(x)}{\lambda q^{(1)}(x)+(1-\lambda)q^{(2)}(x)}-\frac{p^{(1)}(x)}{q^{(1)}(x)}\right|^{n+1}\right. \\ \qquad \left.+(1-\lambda)q^{(2)}(x)\left|\frac{\lambda p^{(1)}(x)+(1-\lambda)p^{(2)}(x)}{\lambda q^{(1)}(x)+(1-\lambda)q^{(2)}(x)}-\frac{p^{(2)}(x)}{q^{(2)}(x)}\right|^{n+1}\right] \\ \frac{1}{n!(n\beta+1)^{\frac{1}{\beta}}}\left\|f^{(n+1)}\right\|_{\alpha}\left[\lambda q^{(1)}(x)\left|\frac{\lambda p^{(1)}(x)+(1-\lambda)p^{(2)}(x)}{\lambda q^{(1)}(x)+(1-\lambda)q^{(2)}(x)}-\frac{p^{(1)}(x)}{q^{(1)}(x)}\right|^{n+\frac{1}{\beta}}\right. \\ \qquad \left.+(1-\lambda)q^{(2)}(x)\left|\frac{\lambda p^{(1)}(x)+(1-\lambda)p^{(2)}(x)}{\lambda q^{(1)}(x)+(1-\lambda)q^{(2)}(x)}-\frac{p^{(2)}(x)}{q^{(2)}(x)}\right|^{n+\frac{1}{\beta}}\right] \\ \frac{1}{n!}\left\|f^{(n+1)}\right\|_{1}\left[\lambda q^{(1)}(x)\left|\frac{\lambda p^{(1)}(x)+(1-\lambda)p^{(2)}(x)}{\lambda q^{(1)}(x)+(1-\lambda)q^{(2)}(x)}-\frac{p^{(1)}(x)}{q^{(1)}(x)}\right|^{n}\right. \\ \qquad \left.+(1-\lambda)q^{(2)}(x)\left|\frac{\lambda p^{(1)}(x)+(1-\lambda)p^{(2)}(x)}{\lambda q^{(1)}(x)+(1-\lambda)q^{(2)}(x)}-\frac{p^{(2)}(x)}{q^{(2)}(x)}\right|^{n}\right], \end{cases}$$

which is equivalent to

$$\begin{aligned} &\Bigg|\left(\lambda q^{(1)}(x)+(1-\lambda)q^{(2)}(x)\right)f\left(\frac{\lambda p^{(1)}(x)+(1-\lambda)p^{(2)}(x)}{\lambda q^{(1)}(x)+(1-\lambda)q^{(2)}(x)}\right) \\ &\qquad -\lambda q^{(1)}(x)f\left(\frac{p^{(1)}(x)}{q^{(1)}(x)}\right)-(1-\lambda)q^{(2)}(x)f\left(\frac{p^{(2)}(x)}{q^{(2)}(x)}\right) \\ &\quad -\lambda\sum_{k=1}^{n}\frac{1}{k!}\frac{(1-\lambda)^{k}(Det_{2,1})^{k}}{\left[\lambda q^{(1)}(x)+(1-\lambda)q^{(2)}(x)\right]^{k}\left[q^{(1)}(x)\right]^{k-1}}f^{(k)}\left(\frac{p^{(1)}(x)}{q^{(1)}(x)}\right) \\ &-(1-\lambda)\sum_{k=1}^{n}\frac{1}{k!}\frac{\lambda^{k}(Det_{1,2})^{k}}{\left[\lambda q^{(1)}(x)+(1-\lambda)q^{(2)}(x)\right]^{k}\left[q^{(2)}(x)\right]^{k-1}}f^{(k)}\left(\frac{p^{(2)}(x)}{q^{(2)}(x)}\right)\Bigg| \end{aligned} \tag{3.16}$$

$$\leq \begin{cases} \frac{1}{(n+1)!}\left\|f^{(n+1)}\right\|_{\infty}\left[\frac{\lambda(1-\lambda)^{n+1}|Det_{2,1}|^{n+1}}{\left[\lambda q^{(1)}(x)+(1-\lambda)q^{(2)}(x)\right]^{n+1}\left[q^{(1)}(x)\right]^{n}}+\frac{(1-\lambda)\lambda^{n+1}|Det_{2,1}|^{n+1}}{\left[\lambda q^{(1)}(x)+(1-\lambda)q^{(2)}(x)\right]^{n+1}\left[q^{(2)}(x)\right]^{n}}\right]; \\ \frac{1}{n!(n\beta+1)^{\frac{1}{\beta}}}\left\|f^{(n+1)}\right\|_{\alpha}\left[\frac{\lambda(1-\lambda)^{n+\frac{1}{\beta}}|Det_{2,1}|^{n+\frac{1}{\beta}}}{\left[\lambda q^{(1)}(x)+(1-\lambda)q^{(2)}(x)\right]^{n+\frac{1}{\beta}}\left[q^{(1)}(x)\right]^{n+\frac{1}{\beta}-1}}\right. \\ \qquad \left.+\frac{(1-\lambda)\lambda^{n+\frac{1}{\beta}}|Det_{2,1}|^{n+\frac{1}{\beta}}}{\left[\lambda q^{(1)}(x)+(1-\lambda)q^{(2)}(x)\right]^{n+\frac{1}{\beta}}\left[q^{(2)}(x)\right]^{n+\frac{1}{\beta}-1}}\right]; \\ \frac{1}{n!}\left\|f^{(n+1)}\right\|_{1}\left[\frac{\lambda(1-\lambda)^{n}|Det_{2,1}|^{n}}{\left[\lambda q^{(1)}(x)+(1-\lambda)q^{(2)}(x)\right]^{n}\left[q^{(1)}(x)\right]^{n-1}}+\frac{(1-\lambda)\lambda^{n}|Det_{2,1}|^{n}}{\left[\lambda q^{(1)}(x)+(1-\lambda)q^{(2)}(x)\right]^{n}\left[q^{(2)}(x)\right]^{n-1}}\right]; \end{cases}$$

$$= \begin{cases} \frac{1}{(n+1)!}\left\|f^{(n+1)}\right\|_{\infty}\frac{\lambda(1-\lambda)|Det_{2,1}|^{n+1}}{\left[\lambda q^{(1)}(x)+(1-\lambda)q^{(2)}(x)\right]^{n+1}}\left[\frac{(1-\lambda)^{n}}{\left[q^{(1)}(x)\right]^{n}}+\frac{\lambda^{n}}{\left[q^{(2)}(x)\right]^{n}}\right]; \\ \frac{1}{n!(n\beta+1)^{\frac{1}{\beta}}}\left\|f^{(n+1)}\right\|_{\alpha}\frac{\lambda(1-\lambda)|Det_{2,1}|^{n+\frac{1}{\beta}}}{\left[\lambda q^{(1)}(x)+(1-\lambda)q^{(2)}(x)\right]^{n+\frac{1}{\beta}}}\left[\frac{(1-\lambda)^{n+\frac{1}{\beta}-1}}{\left[q^{(1)}(x)\right]^{n+\frac{1}{\beta}-1}}+\frac{\lambda^{n+\frac{1}{\beta}-1}}{\left[q^{(2)}(x)\right]^{n+\frac{1}{\beta}-1}}\right]; \\ \frac{1}{n!}\left\|f^{(n+1)}\right\|_{1}\frac{\lambda(1-\lambda)|Det_{2,1}|^{n}}{\left[\lambda q^{(1)}(x)+(1-\lambda)q^{(2)}(x)\right]^{n}}\left[\frac{(1-\lambda)^{n-1}}{\left[q^{(1)}(x)\right]^{n-1}}+\frac{\lambda^{n-1}}{\left[q^{(2)}(x)\right]^{n-1}}\right], \end{cases}$$

where

$$Det_{y,z} = \det \begin{bmatrix} p^{(y)}(x) & p^{(z)}(x) \\ q^{(y)}(x) & q^{(z)}(x) \end{bmatrix}, \qquad (y,z) \in \{(1,2),(2,1)\}.$$

Integrating (3.16) over $x \in \Gamma$ and using the generalised triangle inequality, we deduce the desired inequality (3.12). □

In [17], S.S. Dragomir proved the following perturbed Taylor's formula which is an improvement of a result due to Matić, Pečarić and Ujević [43]. It is instructive to give the details here for the sake of completeness.

Lemma 3.3. *Let $f : I \subseteq \mathbb{R} \to \mathbb{R}$ be such that $f^{(n)}$ is absolutely continuous and $f^{(n+1)} \in L_2(I)$, then we have,*

$$f(z) = \sum_{k=0}^{n} \frac{(z-a)^k}{k!} f^{(k)}(a) + \frac{(z-a)^{n+1}}{(n+1)!}\left[f^{(n)};a,z\right] + G_n(f;a,z) \tag{3.17}$$

for all $z \in I$, where,

$$\left[f^{(n)};a,z\right] = \frac{f^{(n)}(z) - f^{(n)}(a)}{z-a}$$

and $G_n(f;a,z)$ satisfies the estimate,

$$|G_n(f;a,z)| \le \frac{n(z-a)^{n+1}}{(n+1)!\sqrt{2n+1}}\left[\frac{1}{z-a}\left\|f^{(n+1)}\right\|_2^2 - \left(\left[f^{(n)};a,z\right]\right)^2\right]^{\frac{1}{2}} \tag{3.18}$$

for all $z \ge a$.

Proof. Recall Korkine's identity for the mappings h, g,

$$\frac{1}{z-a}\int_a^z h(t)g(t)\,dt - \frac{1}{(z-a)^2}\int_a^z h(t)\,dt \cdot \int_a^z g(t)\,dt = \frac{1}{2(z-a)}\int_a^z\int_a^z (h(t)-h(s))(g(t)-g(s))\,dtds. \tag{3.19}$$

Using (3.19), we have,

$$\int_a^z \frac{(z-t)^n}{n!} f^{(n+1)}(t)\,dt - \frac{1}{z-a}\int_a^z \frac{(z-t)^n}{n!}dt \cdot \int_a^z f^{(n+1)}(t)\,dt = \frac{1}{2(z-a)}\int_a^z\int_a^z \left(\frac{(z-t)^n - (z-s)^n}{n!}\right)\left(f^{(n+1)}(t) - f^{(n+1)}(s)\right)dtds$$

and using Taylor's representation (3.3) and the formula (3.17), we conclude that,

$$G_n(f;a,z) = \frac{1}{2(z-a)}\int_a^z\int_a^z \left[\frac{(z-t)^n - (z-s)^n}{n!}\right]\left(f^{(n+1)}(t) - f^{(n+1)}(s)\right)dtds. \tag{3.20}$$

Now, using the Cauchy-Bunyakovski-Schwarz integral inequality for double integrals, we have,

$$|G_n(f;a,z)| \le \frac{1}{2(z-a)} \left[\int_a^z \int_a^z \left[\frac{(z-t)^n - (z-s)^n}{n!} \right]^2 dtds \right.$$
$$\left. \times \int_a^z \int_a^z \left[f^{(n+1)}(t) - f^{(n+1)}(s) \right]^2 dtds \right]^{\frac{1}{2}}. \quad (3.21)$$

Elementary calculations show that,

$$\frac{1}{2(z-a)^2} \int_a^z \int_a^z \left[\frac{(z-t)^n - (z-s)^n}{n!} \right]^2 dtds = \frac{n^2 (z-a)^{2n}}{[(n+1)!]^2 (2n+1)}$$

and (see also [43])

$$\frac{1}{2(z-a)^2} \int_a^z \int_a^z \left[f^{(n+1)}(t) - f^{(n+1)}(s) \right]^2 dtds = \frac{1}{z-a} \left\| f^{(n+1)} \right\|_2^2 - \left(\left[f^{(n)}; a, z \right] \right)^2,$$

and so, by (3.21), we deduce (3.18). □

Now, by the Grüss inequality, we may state that,

$$0 \le \frac{1}{z-a} \int_a^z \left[f^{(n+1)}(t) \right]^2 dt - \left(\frac{1}{z-a} \int_a^z f^{(n+1)}(t)\, dt \right)^2 \le \frac{1}{4} (\Gamma(z) - \gamma(z)),$$

where

$$\gamma(z) \le f^{(n+1)}(t) \le \Gamma(z) \quad \text{for all } t \in [a,z]. \quad (3.22)$$

By Lemma 3.3, we can obtain the result in [17], showing that (3.18) is an improvement on the pre-Grüss result obtained in [43].

Corollary 3.4. *Let $f : I \subseteq \mathbb{R} \to \mathbb{R}$ be such that $f^{(n)}$ is absolutely continuous and $f^{(n+1)}$ is bounded and satisfies (3.22), then we have the representation (3.17) and the remainder $G_n(f;a,z)$ satisfies the estimate,*

$$|G_n(f;a,z)| \le \frac{n(z-a)^{n+1}}{2(n+1)!\sqrt{2n+1}} (\Gamma(z) - \gamma(z)) \quad (3.23)$$

for all $z \ge a$.

If $z \le a$, then a similar bound can be stated and so, in general, for any $a \in I$, we have the representation (3.17) and the bounds,

$$|G_n(f;a,z)| \le \frac{n|z-a|^{n+1}}{(n+1)!\sqrt{2n+1}} \left[\frac{\int_a^z \left[f^{(n+1)}(t) \right]^2 dt}{z-a} - \left(\left[f^{(n)}; a, z \right] \right)^2 \right]^{\frac{1}{2}} \quad (3.24)$$
$$\le \frac{n|z-a|^{n+1}}{2(n+1)!\sqrt{2n+1}} (\Gamma(z) - \gamma(z)),$$

where

$$\Gamma := \sup_{z \in I} f^{(n+1)}(z) < \infty \quad \text{and} \quad \gamma := \inf_{z \in I} f^{(n+1)}(z) > -\infty.$$

In what follows, we use the estimate (3.24) [3].

Theorem 3.5. *Let $f:(0,\infty)\to\mathbb{R}$ be n-time differentiable and such that $f^{(n)}$ is absolutely continuous on $[r,R]$, where $0<r\le 1\le R<\infty$. Assume that the probability distributions p,q satisfy the condition,*

$$r \le \frac{p(x)}{q(x)} \le R \text{ a.e on } \Gamma, \tag{3.25}$$

then we have the inequalities,

$$\begin{aligned}
&\left| I_f(p,q) - f(1) - \left[\sum_{k=1}^{n} \frac{f^{(k)}(1)}{k!} D_{\chi^k}(p,q)\right] \right. \\
&\qquad \left. - \frac{1}{(n+1)!} I_{(\cdot-1)^k f^{(k)}(\cdot)}(p,q) + \frac{f^{(n)}(1)}{(n+1)!} D_{\chi^n}(p,q) \right| \\
&\le \frac{n}{(n+1)!\sqrt{2n+1}} B\left(p,q,f^{(n+1)}\right) \\
&\le \frac{n(\Phi-\phi)}{(n+1)!\sqrt{2n+1}} D_{|\chi|^{n+1}}(p,q) \le \frac{n(\Phi-\phi)}{(n+1)!\sqrt{2n+1}} (R-r)^{n+1},
\end{aligned} \tag{3.26}$$

where

$$\Phi := \sup_{z\in[r,R]} f^{(n+1)}(z) < \infty \text{ and } \phi := \inf_{z\in[r,R]} f^{(n+1)}(z) > -\infty$$

and

$$B\left(p,q,f^{(n+1)}\right) := I_g(p,q)$$

where

$$g(z) = |z-1|^{n+1} \left[\frac{1}{z-1}\int_1^z \left[f^{(n+1)}(t)\right]^2 dt - \left(\frac{f^{(n)}(z)-f^{(n)}(1)}{z-1}\right)^2\right]^{\frac{1}{2}}.$$

Proof. Apply the inequality (3.24) for $a=1$ and $z=\frac{p(x)}{q(x)}$ to obtain,

$$\left| f\left(\frac{p(x)}{q(x)}\right) - f(1) - \sum_{k=1}^{n} \frac{f^{(k)}(1)}{k!}\left(\frac{p(x)}{q(x)}-1\right)^k - \frac{\left(\frac{p(x)}{q(x)}-1\right)^n}{(n+1)!}\left[f^{(n)}\left(\frac{p(x)}{q(x)}\right) - f^{(n)}(1)\right] \right|$$

$$\leq \frac{n\left|\frac{p(x)}{q(x)}-1\right|^{n}}{(n+1)!\sqrt{2n+1}}\frac{q(x)}{p(x)-q(x)}\int_{1}^{\frac{p(x)}{q(x)}}\left[f^{(n+1)}(t)\right]^{2}dt-\left(\frac{f^{(n)}\left(\frac{p(x)}{q(x)}\right)-f^{(n)}(1)}{\frac{p(x)}{q(x)}-1}\right)^{2}\Bigg]^{\frac{1}{2}}$$

$$\leq \frac{n\left|\frac{p(x)}{q(x)}-1\right|^{n+1}(\Phi-\phi)}{(n+1)!\sqrt{2n+1}}.$$

If we multiply by $p(x)\geq 0$, integrate over $x\in\Gamma$ and use the generalised triangle inequality, we deduce,

$$\Bigg|I_f(p,q)-f(1)-\sum_{k=1}^{n}\frac{f^{(k)}(1)}{k!}D_{\chi^k}(p,q) \\ -\frac{1}{(n+1)!}\int_{\Gamma}q(x)\left(\frac{p(x)}{q(x)}-1\right)^{k}f^{(k)}\left(\frac{p(x)}{q(x)}\right)d\mu(x)+\frac{f^{(n)}(1)}{(n+1)!}D_{\chi^n}(p,q)\Bigg|$$

$$\leq \frac{n(\Phi-\phi)}{(n+1)!\sqrt{2n+1}}\int_{\Gamma}q(x)\left|\frac{p(x)}{q(x)}-1\right|^{n+1}\left[\frac{q(x)}{p(x)-q(x)}\int_{1}^{\frac{p(x)}{q(x)}}\left[f^{(n+1)}(t)\right]^{2}dt \right.\\ \left. -q^{2}(x)\frac{\left|f^{(n)}\left(\frac{p(x)}{q(x)}\right)-f^{(n)}(1)\right|^{2}}{(p(x)-q(x))^{2}}\right]^{\frac{1}{2}}d\mu(x)$$

$$\leq \frac{n(\Phi-\phi)}{2(n+1)!\sqrt{2n+1}}D_{|\chi|^{n+1}}(p,q)\leq \frac{n(\Phi-\phi)}{2(n+1)!\sqrt{2n+1}}(R-r)^{n+1}$$

and the theorem is proved. □

3.2. Some Particular Inequalities

The following proposition holds [3].

Proposition 3.6. *Let p,q be two probability distributions satisfying the condition*

$$0<r\leq\frac{p(x)}{q(x)}\leq R<\infty \text{ a.e on } \Gamma, \tag{3.27}$$

then, for $n\geq 1$, we have the inequality,

$$\left|D_{KL}(q,p)-\sum_{k=1}^{n}\frac{(-1)^k}{k!}D_{\chi^k}(p,q)\right| \tag{3.28}$$

$$\leq\begin{cases}\frac{1}{(n+1)r^{n+1}}D_{|\chi|^{n+1}}(p,q);\\ \frac{1}{(n\beta+1)^{\frac{1}{\beta}}[(n+1)\alpha-1]^{\frac{1}{\alpha}}}\left[\frac{R^{(n+1)\alpha-1}-r^{(n+1)\alpha-1}}{R^{(n+1)\alpha-1}r^{(n+1)\alpha-1}}\right]^{\frac{1}{\alpha}}D_{|\chi|^{n+\frac{1}{\beta}}}(p,q), \quad \alpha>1,\ \frac{1}{\alpha}+\frac{1}{\beta}=1;\\ \frac{1}{n}\cdot\frac{R^n-r^n}{R^n r^n}D_{|\chi|^n}(p,q);\end{cases}$$

$$\leq \begin{cases} \frac{1}{(n+1)r^{n+1}}(R-r)^{n+1}; \\ \frac{1}{(n\beta+1)^{\frac{1}{\beta}}[(n+1)\alpha-1]^{\frac{1}{\alpha}}}\left[\frac{R^{(n+1)\alpha-1}-r^{(n+1)\alpha-1}}{R^{(n+1)\alpha-1}r^{(n+1)\alpha-1}}\right]^{\frac{1}{\alpha}}(R-r)^{n+\frac{1}{\beta}}, & \alpha>1,\ \frac{1}{\alpha}+\frac{1}{\beta}=1; \\ \frac{1}{n}\cdot\frac{R^n-r^n}{R^n r^n}(R-r)^n; \end{cases}$$

Proof. Consider the mapping $f(t) = \ln t$. We have,

$$\begin{aligned} I_f(p,q) &= \int_\Gamma q(x) f\left(\frac{p(x)}{q(x)}\right) d\mu(x) = \int_\Gamma q(x)\ln\left(\frac{p(x)}{q(x)}\right) d\mu(x) \\ &= -\int_\Gamma q(x)\ln\left(\frac{q(x)}{p(x)}\right) d\mu(x) = -D_{KL}(q,p), \end{aligned}$$

$$f^{(k)}(t) = \frac{(-1)^{k-1}(k-1)!}{t^k}, \quad k\in\mathbb{N}, k\geq 1$$

for $\alpha > 1$ and

$$\left\|f^{(n+1)}\right\|_\infty := \sup_{t\in[r,R]}\left|f^{(n+1)}(t)\right| = n!\sup_{t\in[r,R]}\left\{\frac{1}{t^{n+1}}\right\} = \frac{n!}{r^{n+1}};$$

$$\begin{aligned} \left\|f^{(n+1)}\right\|_\alpha &:= \left(\int_r^R \left|f^{(n+1)}(t)\right|^\alpha dt\right)^{\frac{1}{\alpha}} = n!\left[\int_r^R \frac{dt}{t^{(n+1)\alpha}}\right]^{\frac{1}{\alpha}} \\ &= n!\left[\frac{t^{-(n+1)\alpha+1}}{-(n+1)\alpha+1}dt\Bigg|_r^R\right]^{\frac{1}{\alpha}} = n!\left[\frac{R^{(n+1)\alpha-1}-r^{(n+1)\alpha-1}}{[(n+1)\alpha-1]R^{(n+1)\alpha-1}r^{(n+1)\alpha-1}}\right]^{\frac{1}{\alpha}}. \end{aligned}$$

Applying Theorem 3.1 and using the above assumptions, we deduce the desired inequality (3.28). □

The following proposition also holds [3].

Proposition 3.7. *Let p,q be as in the above Proposition 3.6, then we have the inequality,*

$$\left|D_{KL}(q,p) - \sum_{k=2}^n \frac{(-1)^k}{(k-1)k} D_{\chi^k}(p,q)\right| \tag{3.29}$$

$$\leq \begin{cases} \frac{1}{n(n+1)r^{n+1}}D_{|\chi|^{n+1}}(p,q); \\ \frac{1}{n(n\beta+1)^{\frac{1}{\beta}}(n\alpha-1)^{\frac{1}{\alpha}}}\left[\frac{R^{n\alpha-1}-r^{n\alpha-1}}{R^{n\alpha-1}r^{n\alpha-1}}\right]^{\frac{1}{\alpha}} D_{|\chi|^{n+\frac{1}{\beta}}}(p,q), & \alpha>1,\ \frac{1}{\alpha}+\frac{1}{\beta}=1; \\ \frac{1}{(n-1)n}\cdot\frac{R^{n-1}-r^{n-1}}{R^{n-1}r^{n-1}}D_{|\chi|^n}(p,q); \end{cases}$$

$$\leq \begin{cases} \frac{1}{n(n+1)r^{n+1}}(R-r)^{n+1}; \\ \frac{1}{n(n\beta+1)^{\frac{1}{\beta}}(n\alpha-1)^{\frac{1}{\alpha}}}\left[\frac{R^{n\alpha-1}-r^{n\alpha-1}}{R^{n\alpha-1}r^{n\alpha-1}}\right]^{\frac{1}{\alpha}}(R-r)^{n+\frac{1}{\beta}}, & \alpha>1,\ \frac{1}{\alpha}+\frac{1}{\beta}=1; \\ \frac{1}{(n-1)n}\cdot\frac{R^{n-1}-r^{n-1}}{R^{n-1}r^{n-1}}(R-r)^n, \end{cases}$$

for $x \geq a$.

Proof. Consider the mapping $f(t) = t\ln(t)$. We have,

$$I_f(p,q) = \int_\Gamma q(x) f\left(\frac{p(x)}{q(x)}\right) d\mu(x) = \int_\Gamma q(x) \frac{p(x)}{q(x)} \ln\left(\frac{p(x)}{q(x)}\right) d\mu(x)$$
$$= \int_\Gamma p(x) \ln\left(\frac{p(x)}{q(x)}\right) d\mu(x) = D_{KL}(p,q),$$

$$f^{(1)}(t) = \ln t + 1,$$
$$f^{(k)}(t) = \frac{(-1)^k (k-2)!}{t^{k-1}}, \quad k \geq 2$$
$$\left\| f^{(n+1)} \right\|_\infty = \frac{(n-1)!}{r^n},$$
$$\left\| f^{(n+1)} \right\|_\alpha = (n-1)! \left[\frac{R^{n\alpha-1} - r^{n\alpha-1}}{R^{n\alpha-1} r^{n\alpha-1}}\right]^{\frac{1}{\alpha}}, \ \alpha > 1$$

and

$$\left\| f^{(n+1)} \right\|_1 = (n-2)! \frac{R^{n-1} - r^{n-1}}{R^{n-1} r^{n-1}}.$$

Applying Theorem 3.1 for the mapping $f(t) = t\ln t$, we have,

$$\left| D_{KL}(p,q) - f^{(1)}(1) D_\chi(p,q) - \sum_{k=2}^{n} \frac{(-1)^k (k-2)!}{k!} D_{\chi^k}(p,q) \right|$$
$$\leq \begin{cases} \frac{1}{(n+1)!} \frac{(n-1)!}{r^n} D_{|\chi|^{n+1}}(p,q); \\ \frac{1}{n(n\beta+1)^{\frac{1}{\beta}}} \cdot \frac{(n-1)!}{(n\alpha-1)^{\frac{1}{\alpha}}} \left[\frac{R^{n\alpha-1} - r^{n\alpha-1}}{R^{n\alpha-1} r^{n\alpha-1}}\right]^{\frac{1}{\alpha}} D_{|\chi|^{n+\frac{1}{\beta}}}(p,q); \\ \frac{1}{n!}(n-2)! \frac{R^{n-1} - r^{n-1}}{R^{n-1} r^{n-1}} D_{|\chi|^n}(p,q), \end{cases}$$

i.e.,

$$\left| D_{KL}(q,p) - \sum_{k=2}^{n} \frac{(-1)^k}{(k-1)k} D_{\chi^k}(p,q) \right| \leq \begin{cases} \frac{1}{n(n+1) r^{n+1}} D_{|\chi|^{n+1}}(p,q); \\ \frac{1}{n(n\beta+1)^{\frac{1}{\beta}} (n\alpha-1)^{\frac{1}{\alpha}}} \left[\frac{R^{n\alpha-1} - r^{n\alpha-1}}{R^{n\alpha-1} r^{n\alpha-1}}\right]^{\frac{1}{\alpha}} D_{|\chi|^{n+\frac{1}{\beta}}}(p,q); \\ \frac{1}{(n-1)n} \frac{R^{n-1} - r^{n-1}}{R^{n-1} r^{n-1}} D_{|\chi|^n}(p,q) \end{cases}$$

and the first inequality in (3.29) is proved.

The second inequality is obvious and we omit the details. □

Remark 3.1. *Similar results can be obtained if we apply Theorem 3.1 for other particular mappings f, generating the Hellinger, Jeffreys, Bhattacharyya, or other divergence measures as considered in the Introduction.*

4. Approximating Two Mappings Associated with f–Divergence

4.1. Two Functions Associated with f–Divergences

In [2], the authors introduced the following divergence measure

$$H_f(p,q;t) := \int_{\Gamma} p(x) f\left[\frac{tq(x)+(1-t)p(x)}{p(x)}\right] d\mu(x), \tag{4.1}$$

where $p,q \in Q$ and $t \in [0,1]$.

It is obvious that this measure can be represented in terms of f–divergence, namely, we have the representation,

$$H_f(p,q;t) = D_f(p, tq + (1-t)p) \tag{4.2}$$

for all $p,q \in Q$ and $t \in [0,1]$.

The following properties of $H_f(\cdot,\cdot;\cdot)$ hold (see [2]).

Theorem A. *Assume that the mapping $f : [0,\infty) \to \mathbb{R}$ is convex and $p,q \in Q$, then,*

(i) *$H_f(p,q;\cdot)$ is convex on $[0,1]$;*

(ii)

$$H_f(p,q;t) \le D_f(p,q) \quad \textit{for all } t \in [0,1] \tag{4.3}$$

with the bounds

$$\inf_{t\in[0,1]} H_f(p,q;t) = H_f(p,q;0) = 0 \tag{4.4}$$

and

$$\sup_{t\in[0,1]} H_f(p,q;t) = H_f(p,q;1) = D_f(p,q); \tag{4.5}$$

(iii) *The mapping $H_f(p,q;\cdot)$ is monotonic nondecreasing on $[0,1]$.*

In the same paper [2], the authors introduced the following divergence,

$$F_f(p,q;t) = \int_{\Gamma}\int_{\Gamma} p(x)p(y) f\left[t\cdot\frac{q(x)}{p(x)} + (1-t)\cdot\frac{q(y)}{p(y)}\right] d\mu(x)\, d\mu(y), \tag{4.6}$$

where $p,q \in \Omega$ and $t \in [0,1]$.

The properties of this mapping are embodied in the following theorem [2].

Theorem B. *Under the assumptions of Theorem A, we have,*

(i) *$F_f(p,q;\cdot)$ is symmetrical about $\frac{1}{2}$, i.e.,*

$$F_f(p,q;t) = F_f(p,q;1-t) \quad \textit{for all } t \in [0,1]; \tag{4.7}$$

(ii) *$F_f(p,q;\cdot)$ is convex on $[0,1]$;*

(iii) We have the bounds

$$\sup_{t\in[0,1]} F_f(p,q;t) = F_f(p,q;0) = F_f(p,q;1) = D_f(p,q); \tag{4.8}$$

$$\begin{aligned}\inf_{t\in[0,1]} F_f(p,q;t) &= F_f\left(p,q;\frac{1}{2}\right) \\ &= \int_\Gamma\int_\Gamma p(x)p(y)f\left[\frac{q(x)p(y)+p(x)q(y)}{2p(x)p(y)}\right]d\mu(x)\,d\mu(y) \geq 0;\end{aligned} \tag{4.9}$$

(iv) $F_f(p,q;\cdot)$ *is nondecreasing on* $\left[0,\frac{1}{2}\right]$ *and nonincreasing on* $\left[\frac{1}{2},1\right]$*;*

(v) and

$$F_f(p,q;t) \geq \max\left\{H_f(p,q;t), H_f(p,q;1-t)\right\} \tag{4.10}$$

for all $t\in[0,1]$.

In this section we point out some estimates for the divergence measures $F_f(\cdot,\cdot;\cdot)$ and $H_f(\cdot,\cdot;\cdot)$.

4.2. Some Estimates for $n-$Time Differentiable Mappings

We use the following lemma (see also [18]).

Lemma 4.1. *Let* $f: I\in\mathbb{R}\to\mathbb{R}$ *(I interval of* $\mathbb{R}$*) be such that* $f^{(n)}$ *is absolutely continuous on I, then for all* $x,a\in\mathring{I}$ *(*$\mathring{I}$ *is the interior of I) we have the inequality,*

$$\left|f(x)-f(a)-\sum_{k=1}^{n}\frac{(x-a)^k}{k!}f^{(k)}(a)\right|$$
$$\leq\begin{cases}\frac{1}{(n+1)!}\left\|f^{(n+1)}\right\|_\infty |x-a|^{n+1} & \text{if } f^{(n+1)}\in L_\infty(I);\\ \frac{1}{n!(n\beta+1)^{\frac{1}{\beta}}}\left\|f^{(n+1)}\right\|_\alpha |x-a|^{n+\frac{1}{\beta}} & \text{if } f^{(n+1)}\in L_\alpha(I),\\ & \alpha>1,\ \frac{1}{\alpha}+\frac{1}{\beta}=1;\\ \frac{1}{n!}\left\|f^{(n+1)}\right\|_1 |x-a|^n, & \end{cases} \tag{4.11}$$

where $\|\cdot\|_\alpha$ $(\alpha\in[1,\infty])$ *are the usual Lebesgue norms on I, namely,*

$$\|g\|_\alpha := \left(\int_I |g(x)|^\alpha dx\right)^{\frac{1}{\alpha}},\quad \alpha\geq 1$$
$$\|g\|_\infty := ess\sup_{x\in I}|g(x)|.$$

The following corollary will be useful in what follows.

Corollary 4.2. *Assume that f is as above and* $a,b\in\mathring{I}$*, then for all* $\lambda\in[0,1]$*, we have the inequality:*

$$\left|f(\lambda b+(1-\lambda)a)-f(a)-\sum_{k=1}^{n}\frac{\lambda^k(b-a)^k}{k!}f^{(k)}(a)\right|$$

$$\leq \begin{cases} \frac{\lambda^{n+1}|b-a|^{n+1}}{(n+1)!}\left\|f^{(n+1)}\right\|_\infty & \text{if } f^{(n+1)} \in L_\infty(I); \\ \frac{\lambda^{n+\frac{1}{\beta}}|b-a|^{n+\frac{1}{\beta}}}{n!(n\beta+1)^{\frac{1}{\beta}}}\left\|f^{(n+1)}\right\|_\infty & \text{if } f^{(n+1)} \in L_\alpha(I), \\ & \alpha>1,\ \frac{1}{\alpha}+\frac{1}{\beta}=1; \\ \frac{\lambda^n|b-a|^n}{n!}\left\|f^{(n+1)}\right\|_1. \end{cases} \tag{4.12}$$

We can now point out the following estimation result for the mapping $H_f(p,q;\cdot)$ [4].

Theorem 4.3. *Assume that the mapping $f:[0,\infty)\to\mathbb{R}$ is such that $f^{(n)}$ is absolutely continuous on $[r,R]$, where $0\leq r\leq 1\leq R<\infty$. If $p,q\in\Omega$ and*

$$r\leq \frac{q(x)}{p(x)}\leq R \ \text{ a.e. on } \chi, \tag{4.13}$$

then we have,

$$\left|H_f(p,q;t)-f(1)-\sum_{k=1}^{n}\frac{t^k f^{(k)}(1)}{k!}D_{\chi^k}(p,q)\right| \tag{4.14}$$

$$\leq \begin{cases} \frac{t^{n+1}\left\|f^{(n+1)}\right\|_\infty}{(n+1)!}D_{|\chi|^{n+1}}(p,q) & \text{if } f^{(n+1)}\in L_\infty[r,R]; \\ \frac{t^{n+\frac{1}{\beta}}\left\|f^{(n+1)}\right\|_\alpha}{n!(n\beta+1)^{\frac{1}{\beta}}}D_{|\chi|^{n+\frac{1}{\beta}}}(p,q) & \text{if } f^{(n+1)}\in L_\alpha[r,R], \\ & \alpha>1,\ \frac{1}{\alpha}+\frac{1}{\beta}=1; \\ \frac{t^n\left\|f^{(n+1)}\right\|_1}{n!}D_{|\chi|^n}(p,q), \end{cases} \leq \begin{cases} \frac{t^{n+1}(R-r)^{n+1}}{(n+1)!}\left\|f^{(n+1)}\right\|_\infty \\ \frac{t^{n+\frac{1}{\beta}}(R-r)^{n+\frac{1}{\beta}}}{n!(n\beta+1)^{\frac{1}{\beta}}}\left\|f^{(n+1)}\right\|_\alpha \\ \frac{t^n(R-r)^n}{n!}\left\|f^{(n+1)}\right\|_1, \end{cases}$$

where

$$D_{\chi^k}(p,q) := \int_\Gamma \frac{(q(x)-p(x))^k}{p^{k-1}(x)}d\mu(x),\ \ k=1,\dots$$

which is an extension of (1.4) and,

$$D_{|\chi|^r}(p,q) := \int_\Gamma \frac{|q(x)-p(x)|^r}{p^{r-1}(x)}d\mu(x),\ \ r\geq 0$$

where the Lebesgue α−norms are taken on $[r,R]$.

Proof. Apply inequality (4.11) for $\lambda=t\in[0,1]$, $b=\frac{q(x)}{p(x)}$, $x\in\Gamma$ and $a=1$, to get,

$$\left|f\left(t\cdot\frac{q(x)}{p(x)}+(1-t)\right)-f(1)-\sum_{k=1}^{n}\frac{t^k\left(\frac{q(x)}{p(x)}-1\right)^k}{k!}f^{(k)}(1)\right|$$

$$\leq \begin{cases} \frac{t^{n+1}\left|\frac{q(x)}{p(x)}-1\right|^{n+1}}{(n+1)!}\left\|f^{(n+1)}\right\|_\infty; \\ \frac{t^{n+\frac{1}{\beta}}\left|\frac{q(x)}{p(x)}-1\right|^{n+\frac{1}{\beta}}}{n!(n\beta+1)^{\frac{1}{\beta}}}\left\|f^{(n+1)}\right\|_\alpha; \\ \frac{t^n\left|\frac{q(x)}{p(x)}-1\right|^n}{n!}\left\|f^{(n+1)}\right\|_1; \end{cases} \leq \begin{cases} \frac{t^{n+1}(R-r)^{n+1}}{(n+1)!}\left\|f^{(n+1)}\right\|_\infty; \\ \frac{t^{n+\frac{1}{\beta}}(R-r)^{n+\frac{1}{\beta}}}{n!(n\beta+1)^{\frac{1}{\beta}}}\left\|f^{(n+1)}\right\|_\alpha; \\ \frac{t^n(R-r)^n}{n!}\left\|f^{(n+1)}\right\|_1, \end{cases} \tag{4.15}$$

for a.e. $x \in \Gamma$.

If we multiply (4.15) by $p(x) \geq 0$, integrate on Γ and use the properties of the integral, we get,

$$\left| \int_\Gamma p(x) f\left(\frac{tq(x)+(1-t)p(x)}{p(x)}\right) d\mu(x) - f(1) - \sum_{k=1}^{n} \frac{t^k f^{(k)}(1)}{k!} \int_\Gamma \frac{(q(x)-p(x))^k}{p^{k-1}(x)} d\mu(x) \right|$$

$$\leq \begin{cases} \frac{t^{n+1}\|f^{(n+1)}\|_\infty}{(n+1)!} \int_\Gamma \frac{|q(x)-p(x)|^{n+1}}{p^n(x)} d\mu(x); \\ \frac{t^{n+\frac{1}{\beta}}\|f^{(n+1)}\|_\alpha}{n!(n\beta+1)^{\frac{1}{\beta}}} \int_\Gamma \frac{|q(x)-p(x)|^{n+\frac{1}{\beta}}}{p^{n+\frac{1}{\beta}-1}(x)} d\mu(x); \\ \frac{t^n\|f^{(n+1)}\|_1}{n!} \int_\Gamma \frac{|q(x)-p(x)|^n}{p^{n-1}(x)} d\mu(x), \end{cases} \leq \begin{cases} \frac{t^{n+1}(R-r)^{n+1}}{(n+1)!} \|f^{(n+1)}\|_\infty \\ \frac{t^{n+\frac{1}{\beta}}(R-r)^{n+\frac{1}{\beta}}}{n!(n\beta+1)^{\frac{1}{\beta}}} \|f^{(n+1)}\|_\alpha \\ \frac{t^n(R-r)^n}{n!} \|f^{(n+1)}\|_1, \end{cases}$$

and the theorem is proved. □

Remark 4.1. *If $n = 0$, then, basically, for an absolutely continuous mapping $f : [r,R] \subset [0,\infty) \to \mathbb{R}$, we have:*

$$|D_f(p,tq+(1-t)p) - f(1)| \leq \begin{cases} t\|f'\|_\infty D_v(p,q) \\ t^{\frac{1}{\beta}}\|f'\|_\alpha D_{|\alpha|^{\frac{1}{\beta}}}(p,q) \\ \|f'\|_1 \end{cases} \leq \begin{cases} t\|f'\|_\infty (R-r) \\ t^{\frac{1}{\beta}}(R-r)^{\frac{1}{\beta}} \cdot \|f'\|_\alpha \\ \|f'\|_1 \end{cases} \quad (4.16)$$

for all $t \in [0,1]$, where $D_v(p,q) = \int_\Gamma |p(x) - q(x)| d\mu(x)$.
If $n = 1$, and taking into account that $D_\chi(p,q) = 0$, then by (4.14) we get, for the mappings whose derivatives f' are absolutely continuous,

$$|D_f(p,tq+(1-t)p) - f(1)| \leq \begin{cases} \frac{t^2\|f'\|_\infty}{2} D_{\chi^2}(p,q) & \text{if } f' \in L_\infty[r,R] \\ \frac{t^{\frac{\beta+1}{\beta}}\|f'\|_\alpha}{(\beta+1)^{\frac{1}{\beta}}} D_{|\chi|^{\frac{\beta+1}{\beta}}}(p,q) & \text{if } f' \in L_\alpha[r,R] \\ t\|f'\|_1 D_v(p,q) \end{cases} \quad (4.17)$$

for all $t \in [0,1]$.

Of course, if we assume that f is convex and normalised, then the left hand side of both (4.16) and (4.17) will become

$$0 \leq D_f(p,tq+(1-t)p)$$

and the inequalities (4.16) and (4.17) will provide some upper bounds for the mapping $H_f(p,q;t)$, $t \in [0,1]$.

Remark 4.2. *If we assume that f'' is absolutely continuous, then from (4.14) we obtain,*

$$\left| H_f(p,q;t) - f(1) - \frac{t^2}{2} f''(1) D_{\chi^2}(p,q) \right|$$

$$\leq \begin{cases} \frac{t^3\|f'''\|_\infty}{6} D_{|\chi|^3}(p,q) & \text{if } f''' \in L_\infty[r,R]; \\ \frac{t^{\frac{2\beta+1}{\beta}}\|f'''\|_\alpha}{2(2\beta+1)^{\frac{1}{\beta}}} D_{|\chi|^{2+\frac{1}{\beta}}}(p,q) & \text{if } f''' \in L_\alpha[r,R], \quad \alpha>1,\ \frac{1}{\alpha}+\frac{1}{\beta}=1; \\ \frac{t^2\|f'''\|_1}{2} D_{|\chi|^2}(p,q), \end{cases} \tag{4.18}$$

which provides an approximation of $H_f(p,q;t)$ by a quadratic in t whose coefficient is dependent on the χ^2 -distance of p and q.

We also note that Theorem 4.3 contains, as a particular case (for $t=1$), an approximation of the $f-$divergence contained in the following corollary.

Corollary 4.4. *With the assumptions of Theorem 4.3, we have,*

$$\left| D_f(p,q) - f(1) - \sum_{k=1}^{n} \frac{f^{(k)}(1)}{k!} D_{\chi^k}(p,q) \right|$$
$$\leq \begin{cases} \frac{\|f^{(n+1)}\|_\infty}{(n+1)!} D_{|\chi|^{n+1}}(p,q) \\ \frac{\|f^{(n+1)}\|_\alpha}{n!(n\beta+1)^{\frac{1}{\beta}}} D_{|\chi|^{n+\frac{1}{\beta}}}(p,q) \\ \frac{\|f^{(n+1)}\|_1}{n!} D_{|\chi|^n}(p,q) \end{cases} \leq \begin{cases} \frac{\|f^{(n+1)}\|_\infty}{(n+1)!} (R-r)^{n+1} \\ \frac{\|f^{(n+1)}\|_\alpha}{n!(n\beta+1)^{\frac{1}{\beta}}} (R-r)^{n+\frac{1}{\beta}} \\ \frac{\|f^{(n+1)}\|_1}{n!} (R-r)^n. \end{cases} \tag{4.19}$$

We also know that for $t=\frac{1}{2}$, we obtain the generalised Lin-Wong $f-$divergence

$$LW_f(p,q) := D_f\left(p, \frac{1}{2}p + \frac{1}{2}q\right)$$

and so, from (4.14), we may state the following estimation for the Lin-Wong $f-$divergence.

Corollary 4.5. *With the assumptions of Theorem 4.3, we have,*

$$\left| LW_f(p,q) - f(1) - \sum_{k=1}^{n} \frac{t^k f^{(k)}(1)}{2^k k!} D_{\chi^k}(p,q) \right|$$
$$\leq \begin{cases} \frac{\|f^{(n+1)}\|_\infty}{2^{n+1}(n+1)!} D_{|\chi|^{n+1}}(p,q) & \text{if } f^{(n+1)} \in L_\infty[r,R]; \\ \frac{\|f^{(n+1)}\|_\alpha}{2^{n+\frac{1}{\beta}} n!(n\beta+1)^{\frac{1}{\beta}}} D_{|\chi|^{n+\frac{1}{\beta}}}(p,q) & \text{if } f^{(n+1)} \in L_\alpha[r,R], \quad \alpha>1,\ \frac{1}{\alpha}+\frac{1}{\beta}=1; \\ \frac{\|f^{(n+1)}\|_1}{2^n n!} D_{|\chi|^n}(p,q). \end{cases} \tag{4.20}$$

Remark 4.3. *Similar particular cases for $n=0$, $n=1$ and $n=2$ may be stated but we omit the details.*

The following theorem also holds [4].

Theorem 4.6. *Assume that the mapping $f:[0,\infty)\to\mathbb{R}$ is such that $f^{(n)}$ is absolutely continuous on $[r,R]$, where $0\le r\le 1\le R<\infty$. If $p,q\in\Omega$ and*

$$r\le\frac{q(x)}{p(x)}\le R \text{ a.e. on }\Gamma, \tag{4.21}$$

then we have the inequality,

$$\left|F_f(p,q;t)-D_f(p,q)-\sum_{k=1}^{n}\frac{t^k}{k!}D^{(*)}_{f^{(k)}}(p,q)\right| \tag{4.22}$$
$$\le\begin{cases}\frac{t^{n+1}}{(n+1)!}D^{(*)}_{n+1}(p,q)\left\|f^{(n+1)}\right\|_{\infty} & \text{if } f^{(n+1)}\in L_{\infty}[r,R];\\ \frac{t^{n+\frac{1}{\beta}}}{n!(n\beta+1)^{\frac{1}{\beta}}}D^{(*)}_{n+\frac{1}{\beta}}(p,q)\left\|f^{(n+1)}\right\|_{\alpha} & \text{if } f^{(n+1)}\in L_{\alpha}[r,R],\\ & \alpha>1,\ \frac{1}{\alpha}+\frac{1}{\beta}=1;\\ \frac{t^n}{n!}D^{(*)}_{n}(p,q)\left\|f^{(n+1)}\right\|_{1}\end{cases}$$
$$\le\begin{cases}\frac{t^{n+1}}{(n+1)!}(R-r)^{n+1}\left\|f^{(n+1)}\right\|_{\infty} & \text{if } f^{(n+1)}\in L_{\infty}[r,R];\\ \frac{t^{n+\frac{1}{\beta}}}{n!(n\beta+1)^{\frac{1}{\beta}}}(R-r)^{n+\frac{1}{\beta}}\left\|f^{(n+1)}\right\|_{\alpha} & \text{if } f^{(n+1)}\in L_{\alpha}[r,R],\\ & \alpha>1,\ \frac{1}{\alpha}+\frac{1}{\beta}=1;\\ \frac{t^n}{n!}(R-r)^{n}\left\|f^{(n+1)}\right\|_{1},\end{cases}$$

where

$$D^{(*)}_{f^{(k)}}(p,q)=\int_{\Gamma}\int_{\Gamma}\frac{\left(\det\begin{bmatrix}p(y) & q(y)\\ p(x) & q(x)\end{bmatrix}\right)^k}{[p(x)]^{k-1}[p(y)]^{k-1}}f^{(k)}\left(\frac{q(y)}{p(y)}\right)d\mu(x)\,d\mu(y),\quad k=1,\ldots.$$

$$D^{(*)}_{s}(p,q)=\int_{\Gamma}\int_{\Gamma}\frac{\left|\det\begin{bmatrix}p(y) & q(y)\\ p(x) & q(x)\end{bmatrix}\right|^s}{[p(x)]^{s-1}[p(y)]^{s-1}}d\mu(x)\,d\mu(y),\quad s>0$$

and the α−norms are taken on $[r,R]$.

Proof. We choose in Corollary 4.2, $b=\frac{q(x)}{p(x)}$, $a=\frac{q(y)}{p(y)}$, $x,y\in\chi$ to obtain,

$$\left|f\left(t\cdot\frac{q(x)}{p(x)}+(1-t)\cdot\frac{q(y)}{p(y)}\right)-f\left(\frac{q(y)}{p(y)}\right)-\sum_{k=1}^{n}\frac{t^k\left(\frac{q(x)}{p(x)}-\frac{q(y)}{p(y)}\right)^k}{k!}f^{(k)}\left(\frac{q(y)}{p(y)}\right)\right|$$
$$\le\begin{cases}\frac{t^{n+1}}{(n+1)!}\left|\frac{q(x)}{p(x)}-\frac{q(y)}{p(y)}\right|^{n+1}\left\|f^{(n+1)}\right\|_{\infty}\\ \frac{t^{n+\frac{1}{\beta}}}{n!(n\beta+1)^{\frac{1}{\beta}}}\left|\frac{q(x)}{p(x)}-\frac{q(y)}{p(y)}\right|^{n+\frac{1}{\beta}}\left\|f^{(n+1)}\right\|_{\alpha}\\ \frac{t^n}{n!}\left|\frac{q(x)}{p(x)}-\frac{q(y)}{p(y)}\right|^{n}\left\|f^{(n+1)}\right\|_{1}\end{cases}$$

for all $x,y \in \Gamma$ and $t \in [0,1]$, which is clearly equivalent to,

$$\left| f\left(\frac{tp(y)q(x)+(1-t)p(x)q(y)}{p(x)p(y)}\right) - f\left(\frac{q(y)}{p(y)}\right) - \sum_{k=1}^{n} \frac{t^k}{k!} \cdot \frac{\left(\det\begin{bmatrix} p(y) & q(y) \\ p(x) & q(x)\end{bmatrix}\right)^k}{[p(x)]^k [p(y)]^k} f^{(k)}\left(\frac{q(y)}{p(y)}\right) \right|$$

$$\leq \begin{cases} \frac{t^{n+1}}{(n+1)!} \cdot \frac{\left|\det\begin{bmatrix} p(y) & q(y) \\ p(x) & q(x)\end{bmatrix}\right|^{n+1}}{[p(x)]^{n+1}[p(y)]^{n+1}} \left\|f^{(n+1)}\right\|_\infty \\ \frac{t^{n+\frac{1}{\beta}}}{n!(n\beta+1)^{\frac{1}{\beta}}} \cdot \frac{\left|\det\begin{bmatrix} p(y) & q(y) \\ p(x) & q(x)\end{bmatrix}\right|^{n+\frac{1}{\beta}}}{[p(x)]^{n+\frac{1}{\beta}}[p(y)]^{n+\frac{1}{\beta}}} \left\|f^{(n+1)}\right\|_\alpha \\ \frac{t^n}{n!} \cdot \frac{\left|\det\begin{bmatrix} p(y) & q(y) \\ p(x) & q(x)\end{bmatrix}\right|^{n}}{[p(x)]^{n}[p(y)]^{n}} \left\|f^{(n+1)}\right\|_1 \end{cases} \tag{4.23}$$

for all $x,y \in \chi$ and $t \in [0,1]$.

If we multiply (4.23) by $p(x)p(y) \geq 0$ for $x,y \in \Gamma$, integrate over x and y on Γ and use the properties of the integral, we obtain,

$$\left| \int_\Gamma\int_\Gamma p(x)p(y) f\left(\frac{tp(y)q(x)+(1-t)p(x)q(y)}{p(x)p(y)}\right) d\mu(x)\,d\mu(y) - \int_\Gamma\int_\Gamma p(x)p(y) f\left(\frac{q(y)}{p(y)}\right) d\mu(x)\,d\mu(y) - \sum_{k=1}^{n} \frac{t^k}{k!} \int_\Gamma\int_\Gamma \frac{\left(\det\begin{bmatrix} p(y) & q(y) \\ p(x) & q(x)\end{bmatrix}\right)^k}{[p(x)]^{k-1}[p(y)]^{k-1}} f^{(k)}\left(\frac{q(y)}{p(y)}\right) d\mu(x)\,d\mu(y) \right|$$

$$\leq \begin{cases} \frac{t^{n+1}}{(n+1)!} \left\|f^{(n+1)}\right\|_\infty \cdot \int_\Gamma\int_\Gamma \frac{\left|\det\begin{bmatrix} p(y) & q(y) \\ p(x) & q(x)\end{bmatrix}\right|^{n+1}}{[p(x)]^{n}[p(y)]^{n}} d\mu(x)\,d\mu(y) \\ \frac{t^{n+\frac{1}{\beta}}}{n!(n\beta+1)^{\frac{1}{\beta}}} \left\|f^{(n+1)}\right\|_\alpha \cdot \int_\Gamma\int_\Gamma \frac{\left|\det\begin{bmatrix} p(y) & q(y) \\ p(x) & q(x)\end{bmatrix}\right|^{n+\frac{1}{\beta}}}{[p(x)]^{n+\frac{1}{\beta}-1}[p(y)]^{n+\frac{1}{\beta}-1}} d\mu(x)\,d\mu(y) \\ \frac{t^n}{n!} \left\|f^{(n+1)}\right\|_1 \cdot \int_\Gamma\int_\Gamma \frac{\left|\det\begin{bmatrix} p(y) & q(y) \\ p(x) & q(x)\end{bmatrix}\right|^{n}}{[p(x)]^{n-1}[p(y)]^{n-1}} d\mu(x)\,d\mu(y), \end{cases}$$

which is clearly equivalent to the first inequality in (4.22).

The second inequality is obvious by the fact that,

$$\left|\frac{q(x)}{p(x)}-\frac{q(y)}{p(y)}\right| \leq R-r \text{ for all } x,y\in\Gamma.$$

The theorem is thus completely proved. □

5. Approximating $f-$Divergences via a Generalised Taylor Formula

5.1. Representation of the $f-$Divergence

We may state the following representation result which is a reformulation of Theorem 1 in [43]:

Theorem 5.1. *Let $\{S_n(\cdot,\cdot)\}_{n\in\mathbb{N}}$ be a sequence of polynomials of two variables satisfying the condition:*

$$\frac{\partial S_n(t,x)}{\partial t} = S_{n-1}(t,x), \quad S_0(t,x) = 1 \textit{ for } x,t\in\mathbb{R} \textit{ and } n\in\mathbb{N}, \tag{5.1}$$

then we have the identity,

$$f(x) = f(a) + \sum_{k=1}^{n}(-1)^{k+1}\left[S_k(x,x)f^{(k)}(x) - S_k(a,x)f^{(k)}(a)\right] + R_n(f;a,x), \tag{5.2}$$

where

$$R_n(f;a,x) := (-1)^n\int_a^x S_n(t,x)f^{(n+1)}(t)\,dt$$

and $f: I\to\mathbb{R}$ is such that $f^{(n)}$ is absolutely continuous on I.

(1) If in (5.2) we set $S_n(t,x) = \frac{1}{n!}(t-x)^n \quad (n\in\mathbb{N})$, we get the Taylor's identity [18]:

$$f(x) = f(a) + \sum_{k=1}^{n}\frac{(x-a)^k}{k!}f^{(k)}(a) + \frac{1}{n!}\int_a^x (x-t)^n f^{(n+1)}(t)\,dt, \qquad x\in I. \tag{5.3}$$

(2) If in (5.2) we set $S_n(t,x) = \frac{1}{n!}\left(t-\frac{a+x}{2}\right)^n$, $(n\in\mathbb{N})$, then we have the identity [43]:

$$f(x) = f(a) + \sum_{k=1}^{n}\frac{(x-a)^k}{2^k k!}\left[f^{(k)}(a) + (-1)^{k+1}f^{(k)}(x)\right] + \frac{(-1)^n}{n!}\int_a^x\left(x-\frac{a+x}{2}\right)^n f^{(n+1)}(t)\,dt, \quad x\in I. \tag{5.4}$$

(3) If in (5.2) we set $S_n(t,x) = \frac{(x-a)^n}{n!}B_n\left(\frac{t-a}{x-a}\right)$, $n\in\mathbb{N}$, $S_0(t,x)=1$, where B_n denotes the Bernoulli polynomial and $B_n := B_n(0)$ are the Bernoulli numbers, then we have the following representation [43],

$$f(x) = f(a) + \frac{x-a}{2}\left[f'(x)+f'(a)\right] - \sum_{k=1}^{\left[\frac{n}{2}\right]}\frac{(x-a)^{2k}}{(2k)!}B_{2k}\left[f^{(2k)}(x) - f^{(2k)}(a)\right]$$

$$+(-1)^n \frac{(x-a)^n}{n!} \int_a^x B_n\left(\frac{t-a}{x-a}\right) f^{(n+1)}(t)\,dt. \quad (5.5)$$

(4) If in (5.2) we set $S_n(t,x) = \frac{(x-a)^n}{n!} E_n\left(\frac{t-a}{x-a}\right)$, $n \in \mathbb{N}$, $S_0(t,x) = 1$, where $E_n(t)$ denotes the Euler polynomials and B_n are the Bernoulli numbers, then we have the representation [43],

$$f(x) = f(a) + 2\sum_{k=1}^{\left[\frac{n+1}{2}\right]} \frac{(x-a)^{2k-1}\left(4^k-1\right)}{(2k)!} \times B_{2k}\left[f^{(2k-1)}(x) + f^{(2k-1)}(a)\right]$$
$$+(-1)^n \frac{(x-a)^n}{n!} \int_a^x E_n\left(\frac{t-a}{x-a}\right) f^{(n+1)}(t)\,dt. \quad (5.6)$$

We are now able to point out the following representation for the $f-$divergence [21].

Theorem 5.2. *Let $\{S_n(t,z)\}_{n\in\mathbb{N}}$ and $f : I \subseteq \mathbb{R} \to \mathbb{R}$ be as in Theorem 5.1. If $p,q \in \Omega$, then we have the representation,*

$$I_f(p,q) = f(1) + \sum_{k=1}^{n} (-1)^{k+1}\left[I_{S_k(\cdot,\cdot)f^{(k)}(\cdot)}(p,q) - I_{S_k(1,\cdot)f^{(k)}(1)}(p,q)\right] + R_f(p,q), \quad (5.7)$$

where the remainder $R_f(p,q)$ can be given by,

$$R_f(p,q) = (-1)^n \int_\Gamma q(x)\left(\int_1^{\frac{p(x)}{q(x)}} S_n\left(t, \frac{p(x)}{q(x)}\right) f^{(n+1)}(t)\,dt\right) d\mu(x). \quad (5.8)$$

Proof. From the representation (5.2), we may write,

$$f\left(\frac{p(x)}{q(x)}\right) = f(1) + \sum_{k=1}^{n} (-1)^{k+1}\left[S_k\left(\frac{p(x)}{q(x)}, \frac{p(x)}{q(x)}\right) f^{(k)}\left(\frac{p(x)}{q(x)}\right)\right.$$
$$\left. -S_k\left(1, \frac{p(x)}{q(x)}\right) f^{(k)}(1)\right] + (-1)^n \int_1^{\frac{p(x)}{q(x)}} S_n\left(t, \frac{p(x)}{q(x)}\right) f^{(n+1)}(t)\,dt \quad (5.9)$$

for all $x \in \Gamma$.

If we multiply (5.9) by $q(x) \geq 0$ $(x \in \Gamma)$, integrate on Γ and take into account that $\int_\Gamma q(x)\,d\mu(x) = 1$, then we obtain the desired representation (5.7). □

The following particular cases are important in applications.

(1) If we use the representation (5.3), we get,

$$I_f(p,q) = f(1) + \sum_{k=1}^{n} \frac{f^{(k)}(1)}{k!} D_k(p,q)$$
$$+ \frac{1}{n!}\int_\Gamma q(x)\left[\int_1^{\frac{p(x)}{q(x)}} \left(\frac{p(x)}{q(x)} - t\right)^n f^{(n+1)}(t)\,dt\right] d\mu(x) \quad (5.10)$$

for all $p,q \in \Omega$, where,

$$D_k(p,q) := \int_\Gamma q^{-k+1}(x)\,(p(x) - q(x))^k\,d\mu(x), \quad k = 1,\ldots,n.$$

(2) From the identity (5.4),

$$I_f(p,q) = f(1) + \sum_{k=1}^{n} \frac{f^{(k)}(1)}{2^k k!} D_k(p,q) + \sum_{k=1}^{n} \frac{(-1)^{k+1}}{2^k k!} I_{(\cdot-1)^k f^{(k)}(\cdot)}(p,q)$$
$$+ \frac{1}{n!} \int_\Gamma q(x) \left[\int_1^{\frac{p(x)}{q(x)}} \left(1 - \frac{p(x)+q(x)}{q(x)}\right)^n f^{(n+1)}(t)\,dt \right] d\mu(x) \quad (5.11)$$

for all $p,q \in \Omega$.

(3) If we use the identity (5.5), we may obtain:

$$I_f(p,q) = f(1) + \int_\Gamma \left[\frac{p(x)-q(x)}{2}\right] f'\left(\frac{p(x)}{q(x)}\right) d\mu(x)$$
$$+ \sum_{k=1}^{\left[\frac{n}{2}\right]} \frac{f^{(2k)}(1)}{(2k)!} D_{2k}(p,q) - \sum_{k=1}^{\left[\frac{n}{2}\right]} \frac{B_{2k}}{(2k)!} I_{(\cdot-1)^{2k} f^{(2k)}(\cdot)}(p,q)$$
$$+ \frac{(-1)^n}{n!} \int_\Gamma q^{-n+1}(x) (p(x)-q(x))^n$$
$$\times \left[\int_1^{\frac{p(x)}{q(x)}} B_n \left(\frac{t-1}{\frac{p(x)}{q(x)}-1} \right) f^{(n+1)}(t)\,dt \right] d\mu(x) \quad (5.12)$$

for all $p,q \in \Omega$.

(4) Finally, by use of identity (5.6), we may write,

$$I_f(p,q) = f(1) + 2 \sum_{k=1}^{\left[\frac{n+1}{2}\right]} \frac{(4^k-1) B_{2k} f^{(k)}(1)}{(2k)!} D_{2k-1}(p,q)$$
$$+ 2 \sum_{k=1}^{\left[\frac{n+1}{2}\right]} \frac{B_{2k}(4^k-1)}{(2k)!} I_{(\cdot-1)^{2k-1} f^{(2k-1)}}(p,q)$$
$$+ \frac{(-1)^n}{n!} \int_\Gamma q^{-n+1}(x) (p(x)-q(x))^n$$
$$\times \left[\int_1^{\frac{p(x)}{q(x)}} E_n \left(\frac{t-1}{\frac{p(x)}{q(x)}-1} \right) f^{(n+1)}(t)\,dt \right] d\mu(x) \quad (5.13)$$

for all $p,q \in \Omega$.

5.2. Bounds for the Remainder

For $a,b \in \mathbb{R}$, we denote

$$\|f\|_{[a,b],p} := \left| \int_a^b |f(t)|^p\,dt \right|^{\frac{1}{p}} \quad \text{if } p \in [1,\infty)$$

and

$$\|f\|_{[a,b],\infty} := ess \sup_{\substack{t\in[a,b] \\ (t\in[b,a])}} |f(t)|.$$

It is obvious that the order $a<b$ or $a>b$ is irrelevant in the definitions of the above Lebesgue $p-$norms.

The following general theorem involving the estimation of the remainder $R_f(p,q)$ holds [21].

Theorem 5.3. *Assume that $\{S_n(t,x)\}_{n\in\mathbb{N}}$ and f are as in Theorem 5.1. If $p,q\in\Omega$, then we have the inequality,*

$$|R_f(p,q)| \le \begin{cases} \int_\Gamma q(x)\left\|f^{(n+1)}\right\|_{\left[1,\frac{p(x)}{q(x)}\right],\infty} \times \left\|S_n\left(\cdot,\frac{p(x)}{q(x)}\right)\right\|_{\left[1,\frac{p(x)}{q(x)}\right],1} d\mu(x), \\ \int_\Gamma q(x)\left\|f^{(n+1)}\right\|_{\left[1,\frac{p(x)}{q(x)}\right],\alpha} \times \left\|S_n\left(\cdot,\frac{p(x)}{q(x)}\right)\right\|_{\left[1,\frac{p(x)}{q(x)}\right],\beta} d\mu(x), \\ \qquad\qquad \text{if } \alpha>1,\ \frac{1}{\alpha}+\frac{1}{\beta}=1, \\ \int_\Gamma q(x)\left\|f^{(n+1)}\right\|_{\left[1,\frac{p(x)}{q(x)}\right],1} \times \left\|S_n\left(\cdot,\frac{p(x)}{q(x)}\right)\right\|_{\left[1,\frac{p(x)}{q(x)}\right],\infty} d\mu(x). \end{cases} \tag{5.14}$$

Proof. We have that

$$|R_f(p,q)| \le \int_\Gamma q(x)\left|\int_1^{\frac{p(x)}{q(x)}} S_n\left(t,\frac{p(x)}{q(x)}\right) f^{(n+1)}(t)\,dt\right| d\mu(x). \tag{5.15}$$

Now, observe that,

$$\left|\int_1^{\frac{p(x)}{q(x)}} S_n\left(t,\frac{p(x)}{q(x)}\right) f^{(n+1)}(t)\,dt\right| \le \left\|f^{(n+1)}\right\|_{\left[1,\frac{p(x)}{q(x)}\right],\infty} \times \left\|S_n\left(\cdot,\frac{p(x)}{q(x)}\right)\right\|_{\left[1,\frac{p(x)}{q(x)}\right],1} \tag{5.16}$$

and, by Hölder's inequality for $\alpha>1$, $\frac{1}{\alpha}+\frac{1}{\beta}=1$,

$$\begin{aligned} &\left|\int_1^{\frac{p(x)}{q(x)}} S_n\left(t,\frac{p(x)}{q(x)}\right) f^{(n+1)}(t)\,dt\right| \\ &\le \left|\int_1^{\frac{p(x)}{q(x)}} \left|S_n\left(t,\frac{p(x)}{q(x)}\right)\right|^\beta dt\right|^{\frac{1}{\beta}} \times \left|\int_1^{\frac{p(x)}{q(x)}} \left|f^{(n+1)}(t)\right|^\alpha dt\right|^{\frac{1}{\alpha}} \\ &= \left\|S_n\left(\cdot,\frac{p(x)}{q(x)}\right)\right\|_{\left[1,\frac{p(x)}{q(x)}\right],\beta} \times \left\|f^{(n+1)}\right\|_{\left[1,\frac{p(x)}{q(x)}\right],\alpha}. \end{aligned} \tag{5.17}$$

Finally,

$$\left|\int_1^{\frac{p(x)}{q(x)}} S_n\left(t,\frac{p(x)}{q(x)}\right) f^{(n+1)}(t)\,dt\right| \le \left\|f^{(n+1)}\right\|_{\left[1,\frac{p(x)}{q(x)}\right],1} \times \left\|S_n\left(\cdot,\frac{p(x)}{q(x)}\right)\right\|_{\left[1,\frac{p(x)}{q(x)}\right],\infty} \tag{5.18}$$

for all $x\in\Gamma$.

Using (5.15) and (5.16) – (5.18), we deduce (5.14). □

Remark 5.1. *If we assume that* $0 \le r \le \frac{p(x)}{q(x)} \le R < \infty$ *for all* $x \in \Gamma$, *then obviously* $r \le 1 \le R$ *and the right side of the inequality (5.14) may be upper bounded by,*

$$\begin{cases} \left\| f^{(n+1)} \right\|_{[r,R],\infty} \int_\Gamma q(x) \left\| S_n\left(\cdot, \frac{p(x)}{q(x)}\right)\right\|_{\left[1,\frac{p(x)}{q(x)}\right],1} d\mu(x), \\ \left\| f^{(n+1)} \right\|_{[r,R],\alpha} \int_\Gamma q(x) \left\| S_n\left(\cdot, \frac{p(x)}{q(x)}\right)\right\|_{\left[1,\frac{p(x)}{q(x)}\right],\beta} d\mu(x), \\ \left\| f^{(n+1)} \right\|_{[r,R],1} \int_\Gamma q(x) \left\| S_n\left(\cdot, \frac{p(x)}{q(x)}\right)\right\|_{\left[1,\frac{p(x)}{q(x)}\right],\infty} d\mu(x), \end{cases}$$

If we choose some particular instances of polynomials $S_n(\cdot,\cdot)$ we may compute the Lebesgue norm $\left\| S_n\left(\cdot, \frac{p(x)}{q(x)}\right)\right\|_s$, $s \in [1,\infty]$, obtaining more explicit bounds for the remainder $R_f(p,q)$.

(1) If we choose $S_n(t,z) = \frac{1}{n!}(t-z)^n$, then,

$$\|S_n(\cdot,z)\|_{[1,z],1} = \frac{1}{n!}\left|\int_1^z |t-z|^n\, dt\right| = \frac{1}{(n+1)!}|z-1|^{n+1},$$

$$\|S_n(\cdot,z)\|_{[1,z],\alpha} = \frac{1}{n!}\left|\int_1^z |t-z|^{\alpha n}\, dt\right|^{\frac{1}{\alpha}} = \frac{1}{n!}\left[\frac{|z-1|^{\alpha n+1}}{\alpha n+1}\right]^{\frac{1}{\alpha}} = \frac{|z-1|^{n+\frac{1}{\alpha}}}{n!(\alpha n+1)^{\frac{1}{\alpha}}},$$

and

$$\|S_n(\cdot,z)\|_{[1,z],\infty} = \frac{1}{n!}|z-1|^n.$$

Consequently, we may state the following corollary which is useful in practice.

Corollary 5.4. *Let* f *be as in Theorem 5.1, then, for* $p,q \in \Omega$,

$$I_f(p,q) = f(1) + \sum_{k=1}^{n} \frac{f^{(k)}(1)}{k!} D_k(p,q) + R_f(p,q), \tag{5.19}$$

where

$$D_k(p,q) := \int_\Gamma q^{-k+1}(x)(p(x)-q(x))^k\, d\mu(x)$$

and the remainder $R_f(p,q)$ *satisfies the bound*

$$|R_f(p,q)| \tag{5.20}$$

$$\le \begin{cases} \frac{1}{(n+1)!}\int_\Gamma |p(x)-q(x)|^{n+1}[q(x)]^{-n}\left\|f^{(n+1)}\right\|_{\left[1,\frac{p(x)}{q(x)}\right],\infty} d\mu(x), \\ \frac{1}{n!(\alpha n+1)^{\frac{1}{\alpha}}}\int_\Gamma |p(x)-q(x)|^{n+\frac{1}{\alpha}}[q(x)]^{-n-\frac{1}{\alpha}+1}\left\|f^{(n+1)}\right\|_{\left[1,\frac{p(x)}{q(x)}\right],\beta} d\mu(x), \\ \qquad \text{if } \alpha>1,\ \frac{1}{\alpha}+\frac{1}{\beta}=1, \\ \frac{1}{n!}\int_\Gamma |p(x)-q(x)|^{n}[q(x)]^{-n+1}\left\|f^{(n+1)}\right\|_{\left[1,\frac{p(x)}{q(x)}\right],1} d\mu(x). \end{cases}$$

Moreover, if $0 \le r \le \frac{p(x)}{q(x)} \le R < \infty$ *for all* $x \in \Gamma$, *then the right hand side of (5.20) can be upper bounded by,*

$$\begin{cases} \frac{\left\| f^{(n+1)} \right\|_{[r,R],\infty}}{(n+1)!} \int_\Gamma |p(x) - q(x)|^{n+1} [q(x)]^{-n} d\mu(x), \\ \frac{\left\| f^{(n+1)} \right\|_{[r,R],\beta}}{n!(\alpha n+1)^{\frac{1}{\alpha}}} \int_\Gamma |p(x) - q(x)|^{n+\frac{1}{\alpha}} [q(x)]^{-n-\frac{1}{\alpha}+1} d\mu(x), \\ \qquad\qquad \text{if } \alpha > 1,\ \frac{1}{\alpha} + \frac{1}{\beta} = 1, \\ \frac{\left\| f^{(n+1)} \right\|_{[r,R],1}}{n!} \int_\Gamma |p(x) - q(x)|^{n} [q(x)]^{-n+1} d\mu(x), \end{cases} \tag{5.21}$$

$$\le \begin{cases} \frac{\left\| f^{(n+1)} \right\|_{[r,R],\infty} (R-r)^{n+1}}{(n+1)!}, \\ \frac{\left\| f^{(n+1)} \right\|_{[r,R],\beta} (R-r)^{n+\frac{1}{\alpha}}}{n!(\alpha n+1)^{\frac{1}{\alpha}}}, \\ \frac{\left\| f^{(n+1)} \right\|_{[r,R],1} (R-r)^{n}}{n!}. \end{cases}$$

(2) If we choose $S_n(t,z) = \frac{1}{n!}\left(t - \frac{1+z}{2}\right)^n$, then,

$$\|S_n(\cdot,z)\|_{[1,z],1} = \frac{1}{n!}\left| \int_1^z \left| t - \frac{1+z}{2} \right|^n dt \right|.$$

If we assume that $z \ge 1$,

$$\begin{aligned} \int_1^z \left| t - \frac{1+z}{2} \right|^n dt &= \int_1^{\frac{1+z}{2}} \left(\frac{1+z}{2} - t \right)^n dt + \int_{\frac{1+z}{2}}^z \left(t - \frac{1+z}{2} \right)^n dt \\ &= \frac{1}{n+1}\left(\frac{z-1}{2} \right)^{n+1} + \frac{1}{n+1}\left(\frac{z-1}{2} \right)^{n+1} \\ &= \frac{(z-1)^{n+1}}{(n+1)2^n}. \end{aligned}$$

If we assume that $z \le 1$, then,

$$\int_z^1 \left| t - \frac{1+z}{2} \right|^n dt = \frac{(1-z)^{n+1}}{(n+1)2^n}$$

and thus, we may state that,

$$\|S_n(\cdot,z)\|_{[1,z],1} = \frac{1}{(n+1)!} \cdot \frac{|z-1|^{n+1}}{2^n}.$$

Similarly, we have,

$$\begin{aligned} \|S_n(\cdot,z)\|_{[1,z],\alpha} &= \frac{1}{n!}\left| \int_1^z \left| t - \frac{1+z}{2} \right|^{n\alpha} dt \right|^{\frac{1}{\alpha}} = \frac{1}{n!}\left[\frac{|z-1|^{n\alpha+1}}{(n\alpha+1)2^{n\alpha}} \right]^{\frac{1}{\alpha}} \\ &= \frac{1}{n!} \cdot \frac{|z-1|^{n+\frac{1}{\alpha}}}{(n\alpha+1)^{\frac{1}{\alpha}} 2^n}, \quad \alpha \ge 1 \end{aligned}$$

and

$$\|S_n(\cdot,z)\|_{[1,z],\infty}=\frac{1}{n!}\cdot\frac{|z-1|^n}{2^n}.$$

Consequently, we may state the following corollary.

Corollary 5.5. *Let f be as in Theorem 5.1. Then, for $p,q\in\Omega$,*

$$I_f(p,q)=f(1)+\sum_{k=1}^{n}\frac{f^{(k)}(1)}{2^k k!}D_k(p,q)+\sum_{k=1}^{n}\frac{(-1)^{k+1}}{2^k k!}I_{(\cdot-1)^k f^{(k)}(\cdot)}(p,q)+\tilde{R}_f(p,q) \quad (5.22)$$

and the remainder $\tilde{R}_f(p,q)$ satisfies the bound

$$\left|\tilde{R}_f(p,q)\right| \quad (5.23)$$

$$\leq\begin{cases}\frac{1}{(n+1)!2^n}\int_\Gamma[q(x)]^{-n}|p(x)-q(x)|^{n+1}\left\|f^{(n+1)}\right\|_{\left[1,\frac{p(x)}{q(x)}\right],\infty}d\mu(x),\\ \frac{1}{n!2^n(\alpha n+1)^{\frac{1}{\alpha}}}\int_\Gamma[q(x)]^{-n+1-\frac{1}{\alpha}}|p(x)-q(x)|^{n+\frac{1}{\alpha}}\left\|f^{(n+1)}\right\|_{\left[1,\frac{p(x)}{q(x)}\right],\beta}d\mu(x),\\ \qquad\qquad \text{if } \alpha>1,\ \frac{1}{\alpha}+\frac{1}{\beta}=1,\\ \frac{1}{n!2^n}\int_\Gamma[q(x)]^{-n+1}|p(x)-q(x)|^{n}\left\|f^{(n+1)}\right\|_{\left[1,\frac{p(x)}{q(x)}\right],1}d\mu(x).\end{cases}$$

Moreover, if $0\leq r\leq\frac{p(x)}{q(x)}\leq R<\infty$ for all $x\in\Gamma$, then the right hand side of (5.23) can be upper bounded by:

$$\begin{cases}\frac{\left\|f^{(n+1)}\right\|_{[r,R],\infty}}{(n+1)!2^n}\int_\Gamma|p(x)-q(x)|^{n+1}[q(x)]^{-n}d\mu(x),\\ \frac{\left\|f^{(n+1)}\right\|_{[r,R],\beta}}{2^n n!(\alpha n+1)^{\frac{1}{\alpha}}}\int_\Gamma|p(x)-q(x)|^{n+\frac{1}{\alpha}}[q(x)]^{-n-\frac{1}{\alpha}+1}d\mu(x),\\ \qquad\qquad \text{if } \alpha>1,\ \frac{1}{\alpha}+\frac{1}{\beta}=1,\\ \frac{\left\|f^{(n+1)}\right\|_{[r,R],1}}{2^n n!}\int_\Gamma|p(x)-q(x)|^{n}[q(x)]^{-n+1}d\mu(x),\end{cases}$$

$$\leq\begin{cases}\frac{\left\|f^{(n+1)}\right\|_{[r,R],\infty}(R-r)^{n+1}}{(n+1)!2^n},\\ \frac{\left\|f^{(n+1)}\right\|_{[r,R],\beta}(R-r)^{n+\frac{1}{\alpha}}}{n!(\alpha n+1)^{\frac{1}{\alpha}}2^n},\quad \alpha>1,\ \frac{1}{\alpha}+\frac{1}{\beta}=1,\\ \frac{\left\|f^{(n+1)}\right\|_{[r,R],1}(R-r)^{n}}{n!}.\end{cases} \quad (5.24)$$

6. Concluding Remarks

In this survey paper we have shown how Taylor expansions can be used to provide useful approximations of the f-divergence, Shannon's entropy, Kullback-Leibler divergence and other measures of importance in Information Theory.

Bibliography

[1] S.M. Ali and S.D. Silvey, A general class of coefficients of divergence of one distribution from another, *J. Roy. Statist. Soc. Sec B,* **28** (1966), 131–142.

[2] N.S. Barnett, P. Cerone and S.S. Dragomir, Some new inequalities for Hermite-Hadamard difference in information theory, *Stochastic Analysis and Applications*, Y.J. Cho, J.K. Kim and Y.K. Choi (Eds.), Nova Science Publishers, 2003, pp. 7–20.

[3] N.S. Barnett, P. Cerone, S.S. Dragomir and A. Sofo, Approximating Csiszár f−divergence by the use of Taylor's formula with integral remainder, *Math. Ineq. & Appl.,* **5**(3) (2002), 417–434.

[4] N.S. Barnett, P. Cerone, S.S. Dragomir and A. Sofo, Approximating two mappings associated to Csiszár f−divergence via Taylor's expansion, *PanAmerican Math. J.,* **12**(4) (2002), 105-117.

[5] M. Beth Bassat, f-entropies, probability of error and feature selection, *Inform. Control,* **39** (1978), 227–242.

[6] A. Bhattacharyya, On a measure of divergence between two statistical populations defined by their probability distributions, *Bull. Calcutta Math. Soc.,* **35** (1943), 99–109.

[7] I. Burbea and C.R. Rao, On the convexity of some divergence measures based on entropy function, *IEEE Trans. Inf. Th.,* **28**(3) (1982), 489–495.

[8] C.H. Chen, *Statistical Pattern Recognition,* Rocelle Park, New York, Hoyderc Book Co., 1973.

[9] C.K. Chow and C.N. Lin, Approximating discrete probability distributions with dependence trees, *IEEE Trans. Inf. Th.,* **14**(3) (1968), 462–467.

[10] T.M. Cover and J.A. Thomas, *Elements of Information Theory*, John Wiley & Sons Inc., 1991.

[11] I. Csiszár, A note on Jensen's inequality, *Studia Sci. Math. Hung.,* **1** (1966), 185–188.

[12] I. Csiszár, Information-type measures of difference of probability distributions and indirect observations, *Studia Math. Hungarica,* **2** (1967), 299–318.

[13] I. Csiszár, On topological properties of f−divergences, *Studia Math. Hungarica,* **2** (1967), 329–339.

[14] I. Csiszár and J. Körner, *Information Theory: Coding Theorem for Discrete Memoryless Systems,* Academic Press, New York, 1981.

[15] D. Dacunha-Castelle, Ecole d'ete de Probability de Saint-Flour, III-1977, Berlin, Heidelberg: Springer, 1978.

[16] N.M. Dragomir and S.S. Dragomir, An inequality for logarithms and its applications in coding theory, *Indian Journal of Mathematics*, **43**(1) (2001), 13–20.

[17] S.S. Dragomir, An improvement of the remainder estimate in the generalised Taylor's formula, *RGMIA Res. Rep. Coll.,* **3**(1) (2000), Art. 1.

[18] S.S. Dragomir, New estimation of the remainder in Taylor's formula using Grüss' type inequalities and applications, *Math. Ineq. and Appl.*, **2**(2) (1999), 183–193.

[19] S.S. Dragomir, N.M. Dragomir and K. Pranesh, Some estimations of Kraft number and related results, *Soochow Journal of Mathematics*, **24**(4) (1998), 291–296.

[20] S.S. Dragomir and V. Glušč ević, New estimates for the Kullback-Leibler distance and applications, in *Inequality Theory and Applications, Vol. 1*, Y.J. Cho, J.K. Kim and S.S. Dragomir (Eds.), Nova Science Publishers, New York, 2001, pp. 123–138.

[21] S.S. Dragomir and V. Gluščević, Approximating Csiszár f−divergence via a generalised Taylor formula, in *Inequalities for Csiszár f−divergence in Information Theory*, S.S. Dragomir (Ed.), RGMIA Monographs, Victoria University, 2001. [ONLINE:]

[22] S.S. Dragomir and C.J. Goh, A counterpart of Jensen's discrete inequality for differentiable convex mappings and applications in information theory, *Math. Comput. Modelling,* **24**(2) (1996), 1–11.

[23] S.S. Dragomir and C.J. Goh, Some bounds on entropy measures in information theory, *Appl. Math. Lett.,* **10**(3) (1997), 23–28.

[24] S.S. Dragomir, M. Sholz and J. Sŭnde, Some upper bounds for relative entropy and applications, *Computer and Math. With Appl.*, **39**(9-10) (2000), 91–100.

[25] J.H. Havrda and F. Charvat, Quantification method classification process: concept of structural α-entropy, *Kybernetika,* **3** (1967), 30–35.

[26] D.V. Gokhale and S. Kullback, *Information in Contingency Tables*, New York, Merul Dekker, 1978.

[27] E. Hellinger, Neue Bergrüirdung du Theorie quadratisher Formerus von uneudlichvieleu Veränderlicher, *J. für reine and Augeur. Math.,* **36** (1909), 210–271.

[28] H. Jeffreys, An invariant form for the prior probability in estimating problems, *Proc. Roy. Soc. London*, **186** A (1946), 453–461.

[29] T.T. Kadota and L.A. Shepp, On the best finite set of linear observables for discriminating two Gaussian signals, *IEEE Trans. Inf. Th.,* **13** (1967), 288–294.

[30] T. Kailath, The divergence and Bhattacharyya distance measures in signal selection, *IEEE Trans. Comm. Technology.,* Vol COM-15 (1967), 52–60.

[31] J.N. Kapur, A comparative assessment of various measures of directed divergence, *Advances in Management Studies*, **3** (1984), 1-16.

[32] D. Kazakos and T. Cotsidas, A decision theory approach to the approximation of discrete probability densities, *IEEE Trans. Perform. Anal. Machine Intell.,* **1** (1980), 61–67.

[33] J.H.B. Kemperman, On the optimum note of transmitting information, *Ann. Math. Statist.,* **40** (1969), 2158–2177.

[34] C. Kraft, Some conditions for consistency and uniform consistency of statistical procedures, *Univ. of California Pub. in Statistics*, **1** (1955), 125–142.

[35] S. Kullback, A lower bound for discrimination information in terms of variation, *IEEE Trans. Inf. Th.,* **13** (1967), 126–127.

[36] S. Kullback, Correction to a lower bound for discrimination information in terms of variation, *IEEE Trans. Inf. Th.,* **16** (1970), 771–773.

[37] S. Kullback and R.A. Leibler, On information and sufficiency, *Ann. Math. Stat.*, **22** (1951), 79–86.

[38] L. Lecam, *Asymptotic Methods in Statistical Decision Theory,* New York: Springer, 1986.

[39] J. Lin, Divergence measures based on the Shannon entropy, *IEEE Trans. Inf. Th.,* **37** (1) (1991), 145–151.

[40] J. Lin and S.K.M. Wong, A new directed divergence measure and its characterization, *Int. J. General Systems,* **17** (1990), 73–81.

[41] M. Matić, C.E.M. Pearce and J. Pečarić, Improvements of some bounds on entropy measures in information theory, *Math. Inequal. and Appl*., 1 (1998), 295–304.

[42] M. Matić, C.E.M. Pearce and J. Pečarić, Refinements of some bounds in information theory, *ANZIAM J.,* **42**(3) (2001), 387–398.

[43] M. Matić, J. Pečarić and N. Ujević, On new estimation of the remainder in generalised Taylor's formula, *Math. Ineq. and Appl.*, **2**(3) (1999), 343–361.

[44] H.P. McKean, Jr., Speed of approach to equilibrium for Koc's caricature of a Maximilian gas, *Arch. Ration. Mech. Anal.,* **21** (1966), 343–367.

[45] M. Mei, The theory of genetic distance and evaluation of human races, *Japan J. Human Genetics*, **23** (1978), 341–369.

[46] E.C. Pielou, *Ecological Diversity,* Wiley, New York, 1975.

[47] M.S. Pinsker, Information and Information Stability of Random variables and processes, (in Russian), Moscow: Izv. Akad. Nouk, 1960.

[48] C.R. Rao, Diversity and dissimilarity coefficients: a unified approach, *Theoretic Population Biology*, **21** (1982), 24–43.

[49] A. Rényi, On measures of entropy and information, *Proc. Fourth Berkeley Symp. Math. Stat. and Prob., University of California Press*, **1** (1961), 547–561.

[50] A. Sen, *On Economic Inequality,* Oxford University Press, London 1973.

[51] B.D. Sharma and D.P. Mittal, New non-additive measures of relative information, *Journ. Comb. Inf. Sys. Sci.,* **2**(4) (1977), 122–132.

[52] H. Shioya and T. Da-te, A generalisation of Lin divergence and the derivative of a new information divergence, *Elec. and Comm. in Japan*, **78**(7) (1995), 37–40.

[53] I.J. Taneja, *Generalised Information Measures and their Applications* ( ).

[54] H. Theil, *Economics and Information Theory*, North-Holland, Amsterdam, 1967.

[55] H. Theil, *Statistical Decomposition Analysis*, North-Holland, Amsterdam, 1972.

[56] F. Topsøe, Some inequalities for information divergence and related measures of discrimination, *Res. Rep. Coll., RGMIA,* **2**(1) (1999), 85–98.

[57] G.T. Toussaint, Sharper lower bounds for discrimination in terms of variation, *IEEE Trans. Inf. Th.,* **21** (1975), 99–100.

[58] I. Vajda, Note on discrimination information and variation, *IEEE Trans. Inf. Th.,* **16** (1970), 771–773.

[59] I. Vajda, *Theory of Statistical Inference and Information,* Dordrecht-Boston, Kluwer Academic Publishers, 1989.

[60] V.A. Volkonski and J.A. Rozanov, Some limit theorems for random function -*I*, (English Trans.), *Theory Prob. Appl., (*USSR), **4** (1959), 178–197.

In: Advances in Inequalities for Series
Editors: S. S. Dragomir and A. Sofo, pp. 65–88
ISBN 978-1-60021-920-7

Chapter 4

Inequalities for Reciprocals of Power Series with Restricted Coefficients

K. S. Berenhaut[1], *Z. J. Abernathy*[1], *Ying Wai Fan*[2] *and J. D. Foley*[1]
[1]Department of Mathematics
Wake Forest University, Winston-Salem, NC 27106
E-mail addresses: {berenhks, aberzj2, folejd4}@wfu.edu
[2]Department of Mathematics and Computer Science
Emory University, Atlanta, GA 30322
E-mail address: yfan@emory.edu

Abstract

This paper surveys several recent results related to inequalities for coefficients of reciprocals of power series under varying restrictions on the coefficients of the original series. The main techniques used involve obtaining and applying inequalities for linear recurrences with unbounded order. A result for coefficients in intervals that include zero is proven. Some results for multidimensional recurrences and associated multivariable power series are also given. A number of related open questions are included throughout.

1991 Mathematics Subject Classification: 39A10, 39A11, 30B10, 11B37, 65Q05, 06A07.

Key words and phrases: Power Series, Linear Recurrence, Restricted Coefficients, Recurrences with Unbounded Order, Multidimensional recurrences; Multivariable power series; Monotone coefficients.

1. Introduction

For a fixed $I \subset \Re$, let $\mathcal{F}_I$ be the set of I-power series defined by

$$\mathcal{F}_I = \left\{ f : f(z) = 1 + \sum_{k=1}^{\infty} a_k z^k \text{ and } \quad a_k \in I \text{ for each } k \geq 1 \right\}.$$

Flatto, Lagarias, and Poonen [23] and Solomyak [44] proved independently that if z is a complex root of a series in $\mathcal{F}_{[0,1]}$, then $|z| \geq 2/(1+\sqrt{5})$. As $z = -2/(1+\sqrt{5})$ is a root of $1+z+z^3+z^5+\cdots$, this bound is tight over $\mathcal{F}_{[0,1]}$. This result implies that the coefficients of the multiplicative inverse of a series in $\mathcal{F}_{[0,1]}$ cannot increase at a rate larger than the golden ratio. For related results for power series with restricted coefficients cf. Beaucoup *et al.* [1], [2], and Pinner [39]). Related problems for polynomials have been considered by Odlyzko and Poonen [35], Yamamoto [50], Borwein and Pinner [18], and Borwein and Erdelyi [19].

Now, consider computation of the coefficients of the multiplicative inverse h of a series $f \in \mathcal{F}_I$. Equating coefficients in the expansion

$$h_0 + h_1 z + h_2 z^2 + \cdots = \frac{1}{1 + f_1 z + f_2 z^2 + \cdots} \tag{1.1}$$

gives $h_0 = 1$ and

$$h_n = -\sum_{j=0}^{n-1} f_{n-j} h_j, \qquad n \geq 1 \tag{1.2}$$

$$= \sum_{j=0}^{n-1} (-f_{n-j}) h_j. \tag{1.3}$$

The expression in (1.3) suggests consideration of bounds for recurrences with unbounded order. If it is known that $f_i \in -I = \{a : -a \in I\}$ for all $i \geq 1$ then $\{h_i\}$ satisfies a linear recurrence of unbounded order with coefficients in I.

More generally, suppose the sequence $\{b_i\}$ satisfies

$$b_n = \sum_{k=1}^{n-1} \beta_{n,k} b_k, \tag{1.4}$$

for $n \geq 2$, where some limited knowledge is held about the behavior of $\{\beta_{i,j}\}$.

In what follows we will consider several optimal inequalities for recurrences of the type in (1.4) as well as for multidimensional analogues and consider implications of such results for series.

Now, for $V < W$, define the bounding sequence $\{U_i(V,W)\}$ via

$$U_n(V,W) = U_n \stackrel{def}{=} \max\{|b_n| : \{b_i\} \text{ and } \{\beta_{i,j}\} \text{ satisfy (1.4) and } \beta_{i,j} \in [V,W]\}, \tag{1.5}$$

for $n \geq 1$.

In particular, suppose $A > B \geq 0$, and that

$$\beta_{n,k} \in [-A, B], \tag{1.6}$$

for $1 \leq k \leq n-1$ and $n \geq 2$. Without loss of generality we will assume that $b_1 = -1$.

Define the function J via

$$J(n) \stackrel{def}{=} \frac{(1+B)^n - 1}{B}, \tag{1.7}$$

Table 4.1. Values of N^* for various values of A and B

$A\backslash B$	0.001	0.002	0.004	0.008	0.016	0.032	0.064	0.128	0.256	0.512	1.024
0.001	1389										
0.002	813	695									
0.004	449	407	349								
0.008	237	225	205	175							
0.016	123	119	113	103	89						
0.032	63	63	61	59	53	47					
0.064	33	33	33	31	31	27	25				
0.128	17	17	17	17	17	17	15	13			
0.256	9	9	9	9	9	9	9	9	9		
0.512	5	5	5	5	5	5	5	5	5	5	
1.024	3	3	3	3	3	3	3	3	3	3	3
2.048	3	3	3	3	3	3	3	3	3	3	3

for $n \geq 0$. In addition, set

$$N^* = N^*(A,B) \stackrel{def}{=} \max\left\{n : n \text{ is odd and } J(n-3) \leq \frac{2}{A} + \frac{B}{A^2}\right\}.$$

Table 4.1 contains the value of $N^*(A,B)$ for various values of A and B.
In Section 2., the following theorem is proved.

Theorem 1.1. *If $A > B \geq 0$, then*

$$U_n(-A,B) = \begin{cases} A, & \text{if } n=2 \\ \max\{A^2+B, AB+A\}, & \text{if } n=3 \\ A^3 J\left(\left\lfloor \frac{n-1}{2} \right\rfloor\right) J\left(\left\lfloor \frac{n-2}{2} \right\rfloor\right) \\ \qquad +AB[J\left(\left\lfloor \frac{n-1}{2} \right\rfloor\right) + J\left(\left\lfloor \frac{n-2}{2} \right\rfloor\right)] + A, & \text{if } 3 < n < N^* \\ A^2 J(N^*-2) + B, & \text{if } n = N^* \\ AU_{n-1}(-A,B) + (1+B)U_{n-2}(-A,B), & \text{if } n > N^* \end{cases}. \tag{1.8}$$

Recurrences with varying or random coefficients have been studied previously by many authors. A partial survey of such literature contains Viswanath [45] and [46], Viswanath and Trefethen [47], Embree and Trefethen [20], Wright and Trefethen [49], Mallik [29], Popenda [40], Kittapa [27], and Odlyzko [34].

Remark 1.1. *Note that Theorem 1.1 extends a result from [15] where the simpler case $B = 0$ was proven, and compliments the following result from [21].*

Theorem 1.2. *If $A > B \geq 0$ and $A \geq 1$, then,*

$$U_n(-A,-B) = \begin{cases} A, & \text{if } n=2 \\ \max\{A^2-B, A-B^2\}, & \text{if } n=3 \\ A(A^2-2B+1), & \text{if } n=4 \\ A^4-3A^2B+B^2-B, & \text{if } n=5 \\ AU_{n-1}(-A,-B)+(1-B)U_{n-2}(-A,-B), & \text{if } n \geq 6 \end{cases}.$$

Proof. See Theorem 1 in [21]. □

Evaluation of $\{U_i(V,W)\}$ when $|V| < |W|$ with $W > 0$ is trivial, as we then have $U_n(V,W) = W(1+W)^{n-2}$, for $n \geq 2$ (i.e. consider $\beta_{n,k} = W$, for all n and k).

An open question which remains is the following, regarding coefficients in subintervals of the unit interval.

Open Question 1. *For given $0 < D < C < 1$, what is the value of $\{U_n(-C,-D)\}$?*

A solution to this question would complete the pictures for all real intervals. It seems natural to ask whether, in this case, as in Theorems 1.1 and 1.2, $\{U_i\}$ *eventually* satisfies a second order recurrence?

Returning to (1.3), in general, we have the following.

Theorem 1.3. *Suppose that $V < W$, $f \in \mathcal{F}_{[V,W]}$ and f and h satisfy (1.1). Then,*

$$|h_n| \leq U_{n+1}(V,W), \tag{1.9}$$

$n \geq 0$.

Theorem 1.3 may be useful where generating functions or formal power series are utilized such as in enumerative combinatorics and stochastic processes (cf. Wilf [48], Feller [22], Kijima [26], Heathcote [24], Kendall [25]).

The above results provide bounds for the location of the smallest root of a complex valued power series. As mentioned above, Flatto *et al.* [23] and Solomyak [44] independently proved that if z is a root of a series in $\mathcal{F}_{[0,1]}$, then $|z| \geq 2/(1+\sqrt{5})$. The following extension of this result is a consequence of Theorems 1.1 and 1.3.

Corollary 1.4. *Let*

$$\rho = \frac{\delta - A}{2(1+B)}, \tag{1.10}$$

where $\delta = \sqrt{A^2+4(1+B)}$. If z is a root of a power series in $\mathcal{F}_{[-B,A]}$ with $0 \leq B \leq A$, then

$$|z| \geq \rho. \tag{1.11}$$

The result in Corollary 1.4 is optimal: for given $0 \leq B \leq A$,

$$f(z) = 1 + Az - Bz^2 + Az^3 - Bz^4 + Az^5 + \cdots$$

has a root at $z = -\rho$.

Proof (Proof of Corollary 1.4). Suppose that $f \in \mathcal{F}_{[-B,A]}$. Apply Theorems 1.1 and 1.3, and note that $1/f(z)$ is finite for $|z| < \rho$. □

The remainder of the paper proceeds as follows. Section 2. includes some preliminaries and notation, while Section 3. gives a proof of Theorem 1.1. Section 4. surveys some recent results and questions for multivariate power series, while Section 5. deals with results for series with monotone coefficients.

2. Preliminaries and Notation

Suppose $\{b_i\}$ and $\{\beta_{i,j}\}$ satisfy (1.4) and (1.6) with $b_1 = -1$.

Now, let $\mathcal{P} \stackrel{def}{=} \{n \geq 1 : b_n \geq 0\}$ and $\mathcal{N} \stackrel{def}{=} \{n \geq 1 : b_n < 0\}$ partition the sign configuration of $\{b_i\}_{i=1}^{\infty}$, and define B_n (a polynomial in A and B) recursively in n from $\mathcal{N}$ and $\mathcal{P}$ via $B_1 = -1$ and

$$B_n = \left\{ \begin{array}{ll} B \sum\limits_{\substack{1 \leq k \leq n-1 \\ k \in \mathcal{P}}} B_k - A \sum\limits_{\substack{1 \leq k \leq n-1 \\ k \in \mathcal{N}}} B_k, & n \in \mathcal{P} \\ -A \sum\limits_{\substack{1 \leq k \leq n-1 \\ k \in \mathcal{P}}} B_k + B \sum\limits_{\substack{1 \leq k \leq n-1 \\ k \in \mathcal{N}}} B_k, & n \in \mathcal{N} \end{array} \right\}, \tag{2.1}$$

for $n \geq 2$.

A simple induction with (2.1) will show that B_n and b_n always have the same sign.

From the definition, we immediately have [1]

$$|B_n| = A \sum_{\substack{1 \leq k \leq n-1 \\ sign(B_k) \neq sign(B_n)}} |B_k| + B \sum_{\substack{1 \leq k \leq n-1 \\ sign(B_k) = sign(B_n)}} |B_k|, \tag{2.2}$$

for $n \geq 2$.

Note also that the sequence $\{B_k\}$ is an instance of $\{b_k\}$, where the corresponding $\{\beta_{i,j}\}$ are given by

$$\beta_{i,j} = \left\{ \begin{array}{ll} -A, & \text{if } sign(B_i) \neq sign(B_j) \\ B, & \text{if } sign(B_i) = sign(B_j) \end{array} \right., \tag{2.3}$$

for $1 \leq j \leq i-1$ and $i \geq 2$.

The following lemma, reduces our scope from all possible $\{\beta_{i,j}\}$, to the 2^{n-1} various possible sign configurations of $b_2, b_3, \ldots, b_n$. The proof is similar to that of Lemma 1 in [15], and is omitted.

Lemma 2.1. *For all $i \geq 1$,*

$$|b_i| \leq |B_i|.$$

Now, in order to locate the maximal polynomial, B_n (for given n), we will introduce the following notation. For each partition (or two-colouring) of the set $\{1, 2, 3, \ldots, n\}$ into two sets $\mathcal{N}$ and $\mathcal{P}$ with $1 \in \mathcal{N}$, let $\{a_i\}_{i \geq 0}$ denote the "string lengths" of the partition.

[1] We take the $sign(x)$ to be 1 if $x \geq 0$ and -1 otherwise.

For example, suppose $n = 9$, $\mathcal{N} = \{1,2,5,6,7\}$ and $\mathcal{P} = \{3,4,8,9\}$. The sequence of string lengths would then be $(a_0,a_1,a_2,a_3) = (2,2,3,2)$ since $1,2 \in \mathcal{N}$, $3,4 \in \mathcal{P}$, $5,6,7 \in \mathcal{N}$, and $8,9 \in \mathcal{P}$. Thought of as conveying the sign structure of $\{b_n\}$, these particular $\mathcal{N}$ and $\mathcal{P}$ would correspond to $(-,-,+,+,-,-,-,+,+)$. Thus, B_n is completely determined by the sign configuration of $b_1,b_2,\ldots,b_n$, and we can regard B_n as a function on the subset of all possible consecutive "string lengths" where the string lengths sum to n. Specifically, we have

$$B_n = B_n(\langle a_0,a_1,\ldots,a_k\rangle) \text{ where } a_i \geq 1 \text{ and } a_0+a_1+\cdots+a_k = n. \tag{2.4}$$

The following two lemmas follow from the definition in (2.1).

Lemma 2.2. *If $a_0 > 1$, then for $n \geq 3$,*

$$B_n(\langle a_0,a_1,\ldots\rangle) = (1+B)B_{n-1}(\langle a_0-1,a_1,\ldots\rangle). \tag{2.5}$$

Lemma 2.3. *If $a_k \geq 1$, then for all $n \geq 2$,*

$$B_n(\langle a_0,a_1,\ldots,a_k+1\rangle) = (1+B)B_{n-1}(\langle a_0,a_1,\ldots,a_k\rangle). \tag{2.6}$$

Applying Lemmas 2.2 and 2.3 gives the following recursive inequality for $\{U_i\}$.

Corollary 2.4. *For all $n \geq 1$,*

$$U_{n+1} \geq (1+B)U_n. \tag{2.7}$$

In addition, we have the following two results regarding a_0 and a_k.

Lemma 2.5. *For all $n > 1$, there exists $a_1,\ldots,a_k$ such that*

$$U_n = |B_n(\langle 1,a_1,\ldots,a_k\rangle)|.$$

Proof. The result is trivial when $n \leq 2$. Now, suppose $U_{n-1} = B_{n-1}(\langle 1,\hat{a}_1,\ldots,\hat{a}_l\rangle)$, for some $n \geq 2$ and $l \geq 1$. Then, if $a_0 > 1$,

$$\begin{aligned} |B_n(\langle a_0,a_1,\ldots\rangle)| &= (1+B)|B_{n-1}(\langle a_0-1,a_1,\ldots\rangle)| \\ &\leq (1+B)|B_{n-1}(\langle 1,\hat{a}_1,\ldots,\hat{a}_l\rangle)| \\ &= |B_n(\langle 1,\hat{a}_1,\ldots,\hat{a}_l+1\rangle)|, \end{aligned} \tag{2.8}$$

by Lemmas 2.2 and 2.3. □

Lemma 2.6. *For all $n \geq 3$, there exists $a_1,\ldots,a_{k-1}$ such that*

$$U_n = |B_n(\langle 1,a_1,\ldots,a_{k-1},1\rangle)|.$$

Proof. Set $P = \sum_{\substack{1\leq i\leq n-2 \\ i\in\mathcal{P}}} |B_i|$ and $N = \sum_{\substack{1\leq i\leq n-2 \\ i\in\mathcal{N}}} |B_i|$. Then, $N \geq |B_1| = 1 > 0$ and $P > |B_2| > 0$. Now, consider the four cases of $(sign(B_{n-1}), sign(B_n))$ and the associated values of $|B_n|$ (from (2.1)):

(1) $(-,-)$: $|B_n| = AP + BN + ABP + B^2N$,

(2) $(+,-)$: $|B_n| = AP + BN + ABP + A^2N$,

(3) $(+,+)$: $|B_n| = AN + BP + ABN + B^2P$, and

(4) $(-,+)$: $|B_n| = AN + BP + ABN + A^2P$.

Comparing Case 1 with Case 2, and Case 3 with Case 4, and noting that $A > B$, gives the result. □

Considering the result in Lemma 2.6, we define a sequence of functions $\{\phi_j\}$ via

$$\phi_0 = |B_2(\langle 1,1\rangle)| = A, \tag{2.9}$$

and for $j \geq 1$,

$$\phi_j(a_1,a_2,\ldots,a_j) = |B_n(\langle 1,a_1,a_2,\ldots,a_j,1\rangle)|, \tag{2.10}$$

where $\sum_{i=1}^{j} a_j = n-2$.

The problem of finding a bound on $|b_n|$ has been reduced to bounding $|B_n|$ and equivalently to bounding the ϕ_j's over integer sequences $\{a_i\}$ satisfying $\sum_{i=1}^{j} a_i = n-2$.

In what follows, it will be convenient to have the following properties of J, which are immediate from the definition in (1.7).

(1) For positive integers a and b,

$$J(a+b) = J(a) + J(b) + BJ(a)J(b) = J(a) + [1 + BJ(a)]J(b) \tag{2.11}$$

(2) In particular, with $a = 1$, $J(b+1) = 1 + (1+B)J(b)$;

(3) $J(0) = 0$.

We also have the following second order relation for $\{\phi_i\}$, which follows from the definitions in (2.1) and (2.10), and repeated application of Lemma 2.3.

Lemma 2.7. *If $a_i \geq 1$, for $i \geq 1$, then we have*

$$\phi_1(a_1) = A^2 J(a_1) + B, \tag{2.12}$$

$$\phi_2(a_1,a_2) = A^3 J(a_1)J(a_2) + AB[J(a_1) + J(a_2)] + A, \tag{2.13}$$

$$\phi_j(a_1,a_2,\ldots,a_j) = AJ(a_j)\phi_{j-1}(a_1,a_2,\ldots,a_{j-1}) + [1 + BJ(a_{j-1})]\phi_{j-2}(a_1,a_2,\ldots,a_{j-2}), \qquad \textit{for } j \geq 2. \tag{2.14}$$

For convenience, we will at times use ϕ_j to denote $\phi_j(a_1,a_2,\ldots,a_j)$. If $i < j$ arguments to ϕ_j are given, then the i values are for $a_{j-i+1}, a_{j-i+2}, \ldots, a_j$. For instance $\phi_j(b,c)$ may refer to $\phi_j(a_1,a_2,\ldots,a_{j-2},b,c)$.

We now turn to the proof of Theorem 1.1.

3. Proof of Theorem 1.1

Prior to proving Theorem 1.1, we quote the following result from [5] which resolves the form of U_i for small i.

Proposition 3.1. *Suppose $n > 3$ and $J(n-3) \leq \frac{2}{A} + \frac{B}{A^2}$, then*

$$U_n = \begin{cases} \phi_1(N^* - 2), & \text{if } n = N^* \\ \phi_2\left(\lfloor \frac{n-1}{2} \rfloor, \lfloor \frac{n-2}{2} \rfloor\right), & \text{otherwise} \end{cases}$$
$$= \begin{cases} A^2 J(N^*-2) + B, & \text{if } n = N^* \\ A^3 J\left(\lfloor \frac{n-1}{2} \rfloor\right) J\left(\lfloor \frac{n-2}{2} \rfloor\right) \\ \quad + AB\left[J\left(\lfloor \frac{n-1}{2} \rfloor\right) + J\left(\lfloor \frac{n-2}{2} \rfloor\right)\right] + A, & \text{otherwise} \end{cases}.$$

Proof. See [5]. □

We are now in a position to prove our main result on the structure of U_i for large i.

Theorem 3.2. *If $n > N^*$, then*

$$U_n = AU_{n-1} + (1+B)U_{n-2}. \tag{3.1}$$

First, we prove the following recurrence inequality.

Theorem 3.3. *If $J(n-3) \geq \frac{2}{A} + \frac{B}{A^2}$,*

$$U_n \leq AU_{n-1} + (1+B)U_{n-2}. \tag{3.2}$$

Proof. We will consider the following four cases.

(1) $k = 1$;

(2) $k > 1$ and $a_k = 1$;

(3) $k > 1$, $a_k > 1$ and $a_{k-1} = 1$;

(4) $k > 1$, $a_k > 1$ and $a_{k-1} > 1$.

The expansions in (2.14) and (2.11) will be used throughout the proof.

Case 1. $(k = 1)$. *Here $a_1 = n - 2$. Set*

$$S = \phi_k - A\phi_1(n-3) - (1+B)\phi_1(n-4).$$

Then, straightforward applications of Lemma 2.7 and (2.11) give

$$\begin{aligned} S &= \phi_1(n-2) - A\phi_1(n-3) - (1+B)\phi_1(n-4) \\ &= A^2J(n-2) + B - A[A^2J(n-3)+B] - (1+B)[A^2J(n-4)+B] \\ &= A^2[J(n-2) - AJ(n-3) - (1+B)J(n-4)] + B - AB - (1+B)B \\ &= A^2[J(n-3) + BJ(n-3) + 1 - AJ(n-3) - (1+B)J(n-4)] - AB - B^2 \\ &= A^2[J(n-3) - (1+B)J(n-4) + (B-A)J(n-3)] + A^2 - AB - B^2 \\ &= A^2[1 + (B-A)J(n-3)] + A^2 - AB - B^2 \\ &= A^2(B-A)J(n-3) + 2A^2 - AB - B^2 \\ &= A^2(B-A)J(n-3) + (2A+B)(A-B). \end{aligned} \tag{3.3}$$

By the assumption on $J(n-3)$, we then have

$$\begin{aligned} S &= (A-B)[-A^2J(n-3) + (2A+B)] \\ &\le (A-B)\left[-A^2\left(\frac{2}{A}+\frac{B}{A^2}\right) + (2A+B)\right] = 0. \end{aligned} \tag{3.4}$$

Thus, by the definition of $\{U_i\}$, (3.4) gives

$$\phi_k \le A\phi_1(n-3) + (1+B)\phi_1(n-4) \le AU_{n-1} + (1+B)U_{n-2}.$$

Case 2. $(k>1, a_k=1)$. *Note that $J(a_k) = J(1) = 1$. We have*

$$\begin{aligned} \phi_k &= AJ(a_k)\phi_{k-1} + [1 + BJ(a_{k-1})]\phi_{k-2} \\ &= A\phi_{k-1} + (1+B)^{a_{k-1}}\phi_{k-2} \\ &\le AU_{n-1} + (1+B)^{a_{k-1}}U_{n-a_{k-1}-1} \\ &\le AU_{n-1} + (1+B)^{a_{k-1}}\frac{U_{n-2}}{(1+B)^{a_{k-1}-1}} \\ &= AU_{n-1} + (1+B)U_{n-2}, \end{aligned} \tag{3.5}$$

where the last inequality in (3.5) follows from the inequality of Corollary 2.4.

Case 3. $(k>1,\ a_k>1$ *and* $a_{k-1}=1)$. *Note that $J(a_{k-1}) = J(1) = 1$. We have*

$$\begin{aligned} \phi_k &= AJ(a_k)\phi_{k-1} + [1 + BJ(a_{k-1})]\phi_{k-2} \\ &= AJ(a_k)\left\{AJ(a_{k-1})\phi_{k-2} + [1+BJ(a_{k-2})]\phi_{k-3}\right\} + [1+BJ(a_{k-1})]\phi_{k-2} \\ &= A^2J(a_k)J(a_{k-1})\phi_{k-2} + AJ(a_k)[1+BJ(a_{k-2})]\phi_{k-3} + [1+BJ(a_{k-1})]\phi_{k-2}, \end{aligned} \tag{3.6}$$

and

$$\begin{aligned} A\phi_{k-1}(a_k) &= A\{AJ(a_k)\phi_{k-2} + [1+BJ(a_{k-2})]\phi_{k-3}\} \\ &= A^2J(a_k)J(a_{k-1})\phi_{k-2} + A[1+BJ(a_{k-2})]\phi_{k-3}\}. \end{aligned} \tag{3.7}$$

Hence, taking the difference of (3.6) and (3.7), gives

$$\begin{aligned}\phi_k - A\phi_{k-1}(a_k) &= A[J(a_k)-1][1+BJ(a_{k-2})]\phi_{k-3} + (1+BJ(a_{k-1}))\phi_{k-2}\\ &= A[(1+B)J(a_k-1)][1+BJ(a_{k-2})]\phi_{k-3} + (1+B)\phi_{k-2}\\ &= A(1+B)J(a_k-1)[1+BJ(a_{k-2})]\phi_{k-3}\\ &\qquad + (1+B)\{AJ(a_{k-2})\phi_{k-3} + [1+BJ(a_{k-3})]\phi_{k-4}\}\\ &= (1+B)\{A\{J(a_k-1)[1+BJ(a_{k-2})] + J(a_{k-2})\}\phi_{k-3}\\ &\qquad + [1+BJ(a_{k-3})]\phi_{k-4}\}\\ &= (1+B)\{AJ(a_{k-2}+a_k-1)\phi_{k-3} + [1+BJ(a_{k-3})]\phi_{k-4}\}\\ &= (1+B)\phi_{k-2}(a_{k-2}+a_k-1).\end{aligned}$$

Thus,

$$\phi_k = A\phi_{k-1}(a_k) + (1+B)\phi_{k-2}(a_{k-2}+a_k-1) \leq AU_{n-1} + (1+B)U_{n-2}. \tag{3.8}$$

Case 4. ($k > 1$, $a_k > 1$ *and* $a_{k-1} > 1$). *Here, expanding via (3.6) gives*

$$\begin{aligned}\phi_k &= A^2J(a_k)J(a_{k-1})\phi_{k-2} + AJ(a_k)[1+BJ(a_{k-2})]\phi_{k-3} + [1+BJ(a_{k-1})]\phi_{k-2} \qquad (3.9)\\ &= \frac{A^2}{1+B}[J(a_k)J(a_{k-1}) + BJ(a_k)J(a_{k-1})]\phi_{k-2}\\ &\quad + AJ(a_k)[1+BJ(a_{k-2})]\phi_{k-3} + [1+BJ(a_{k-1})]\phi_{k-2}\end{aligned}$$

and

$$\begin{aligned}&A\phi_{k-1}(a_k+a_{k-1}-1) \qquad (3.10)\\ &= A^2J(a_k+a_{k-1}-1)\phi_{k-2} + A[1+BJ(a_{k-2})]\phi_{k-3}\\ &= A^2\left(\frac{J(a_k+a_{k-1})-1}{1+B}\right)\phi_{k-2} + A[1+BJ(a_{k-2})]\phi_{k-3}\\ &= \frac{A^2}{1+B}[J(a_k)+J(a_{k-1}) + BJ(a_k)J(a_{k-1}) - 1]\phi_{k-2} + A[1+BJ(a_{k-2})]\phi_{k-3}.\end{aligned}$$

Thus, combining (3.9) and (3.10) gives

$$\begin{aligned}
&\phi_k - A\phi_{k-1}(a_k + a_{k-1} - 1) \\
&= \frac{A^2}{1+B}\left[J(a_k)J(a_{k-1}) - J(a_k) - J(a_{k-1}) + 1\right]\phi_{k-2} \\
&\qquad + A[J(a_k)-1][1+BJ(a_{k-2})]\phi_{k-3} + [1+BJ(a_{k-1})]\phi_{k-2} \\
&= \frac{A^2}{1+B}\left[J(a_k)-1\right]\left[J(a_{k-1})-1\right]\phi_{k-2} \\
&\qquad + A[J(a_k)-1][1+BJ(a_{k-2})]\phi_{k-3} + [1+BJ(a_{k-1})]\phi_{k-2} \\
&= A[J(a_k)-1]\left\{A\left[\frac{J(a_{k-1})-1}{1+B}\right]\phi_{k-2} + [1+BJ(a_{k-2})]\phi_{k-3}\right\} + [1+BJ(a_{k-1})]\phi_{k-2} \\
&= A(1+B)J(a_k-1)\left\{AJ(a_{k-1}-1)\phi_{k-2} + [1+BJ(a_{k-2})]\phi_{k-3}\right\} \\
&\qquad + [1+B+B(1+B)J(a_{k-1}-1)]\phi_{k-2} \\
&= A(1+B)J(a_k-1)\phi_{k-1}(a_{k-1}-1) + (1+B)[1+BJ(a_{k-1}-1)]\phi_{k-2} \\
&= (1+B)\left\{AJ(a_k-1)\phi_{k-1}(a_{k-1}-1) + [1+BJ(a_{k-1}-1)]\phi_{k-2}\right\} \\
&= (1+B)\phi_k(a_k-1, a_{k-1}-1).
\end{aligned}$$

Finally,

$$\begin{aligned}
\phi_k &= A\phi_{k-1}(a_k + a_{k-1} - 1) + (1+B)\phi_k(a_k-1, a_{k-1}-1) \\
&\le AU_{n-1} + (1+B)U_{n-2}.
\end{aligned} \tag{3.11}$$

□

We will show that (3.2) is in fact an equality by determining $\{a_i\}$ for which ϕ_j attains the optimal bounds.

First, define $\{G_i\}_{i=-2}^{\infty}$ via

$$G_{-2} = \phi_2\left(\frac{N^*-3}{2}, \frac{N^*-3}{2}\right) \tag{3.12}$$

$$G_{-1} = \phi_1(N^*-2) \tag{3.13}$$

$$G_0 = \phi_2\left(\frac{N^*-1}{2}, \frac{N^*-1}{2}\right) \tag{3.14}$$

$$G_i = \phi_{2+i}\left(\frac{N^*-1}{2}, \underbrace{1, 1, \ldots, 1}_{i \text{ times}}, \frac{N^*-1}{2}\right), \quad \text{for } i > 0. \tag{3.15}$$

It is easy to see that each G_i is of the form $\phi_{j_i}(a_{i,1}, a_{i,2}, \ldots, a_{i,j_i})$ with

$$\sum_{l=1}^{j_i} a_{i,l} = i + N^* - 1. \tag{3.16}$$

We have the following.

Lemma 3.4. *For* $i \geq 0$,

$$G_i = AG_{i-1} + (1+B)G_{i-2}, \tag{3.17}$$

for $i \geq 0$.

Proof. First, employing (3.11) and simplifying gives,

$$\begin{aligned} G_0 &= \phi_2\left(\frac{N^*-1}{2}, \frac{N^*-1}{2}\right) \\ &= A\phi_1\left(\frac{N^*-1}{2} + \frac{N^*-1}{2} - 1\right) + (1+B)\phi_2\left(\frac{N^*-1}{2} - 1, \frac{N^*-1}{2} - 1\right) \\ &= A\phi_1(N^*-2) + (1+B)\phi_2\left(\frac{N^*-3}{2}, \frac{N^*-3}{2}\right) \\ &= AG_{-1} + (1+B)G_{-2}. \end{aligned}$$

In addition, applying the identity in (3.8),

$$\begin{aligned} G_1 &= \phi_3\left(\frac{N^*-1}{2}, 1, \frac{N^*-1}{2}\right) \\ &= A\phi_2\left(\frac{N^*-1}{2}, \frac{N^*-1}{2}\right) + (1+B)\phi_1\left(\frac{N^*-1}{2} + \frac{N^*-1}{2} - 1\right) \quad \text{(by (3.8))} \\ &= A\phi_2\left(\frac{N^*-1}{2}, \frac{N^*-1}{2}\right) + (1+B)\phi_1(N^*-2) \\ &= AG_0 + (1+B)G_{-1}. \end{aligned}$$

Similarly, for $i \geq 2$, we have

$$\begin{aligned} G_i &= \phi_{2+i}\left(\frac{N^*-1}{2}, \underbrace{1, 1, \ldots, 1}_{i \text{ times}}, \frac{N^*-1}{2}\right) \\ &= A\phi_{2+i-1}\left(\frac{N^*-1}{2}, \underbrace{1, 1, \ldots, 1}_{(i-1) \text{ times}}, \frac{N^*-1}{2}\right) \\ &\quad + (1+B)\phi_{2+i-2}\left(\frac{N^*-1}{2}, \underbrace{1, 1, \ldots, 1}_{(i-2) \text{ times}}, 1 + \frac{N^*-1}{2} - 1\right) \\ &= AG_{i-1} + (1+B)G_{i-2}, \end{aligned} \tag{3.18}$$

and the lemma is proven. □

We can now prove the eventual second order structure of $\{U_i\}$.

Proof (Proof of Theorem 3.2). By Proposition 3.1, (3.12) and (3.13), we have

$$\begin{aligned} U_{N^*-1} &= \phi_2\left(\frac{N^*-3}{2}, \frac{N^*-3}{2}\right) = G_{-2}, \text{ and} \\ U_{N^*} &= \phi_1(N^*-2) = G_{-1}. \end{aligned} \tag{3.19}$$

In addition, if $N^*+1 \in \left\{n : J(n-3) \leq \frac{2}{A} + \frac{B}{A^2}\right\}$, then by Proposition 3.1, we have

$$U_{N^*+1} = \phi_2\left(\frac{N^*-1}{2}, \frac{N^*-1}{2}\right) = G_0.$$

Otherwise, by Theorem 3.3 and Lemma 3.4,

$$\begin{aligned} U_{N^*+1} &\leq AU_{N^*} + (1+B)U_{N^*-1} \\ &= AG_{-1} + (1+B)G_{-2} \\ &= G_0 \leq U_{N^*+1}. \end{aligned}$$

Hence, we have

$$U_{N^*+1} = G_0 = AG_{-1} + (1+B)G_{-2} = AU_{N^*} + (1+B)U_{N^*-1}.$$

If $n > N^*+1$ then $n \in \left\{n : J(n-3) > \frac{2}{A} + \frac{B}{A^2}\right\}$. Thus, suppose $U_k = G_{k-N^*-1}$ for $N^* - 1 \leq k \leq n-1$. Then, by Theorem 3.3 and Lemma 3.4,

$$\begin{aligned} U_n &\leq AU_{n-1} + (1+B)U_{n-2} \\ &= AG_{n-N^*-2} + (1+B)G_{n-N^*-3} \\ &= G_{n-N^*-1} \leq U_n. \end{aligned}$$

This implies

$$U_n = G_{n-N^*-1} = AU_{n-1} + (1+B)U_{n-2},$$

and the result follows. □

Theorem 1.1 now follows from Proposition 3.1 and Theorem 3.2.

Table 4.2 gives optimal polynomials ϕ_i for some i and various ranges of A (in terms of $B < A$).

Consideration of properties of optimal polynomials arising from $\{U_n(V,W)\}$ appears interesting in its own right (see [15] for further discussion in the case $W = 0$). In this case, the polynomials can be viewed as measuring the number of paths through the sign configuration space with alternating sign steps weighted by B and same sign steps weighted by A (see (2.2)). In [9], more general polynomials are considered as paths through networks and these ideas have since been applied to multivariate power series. Such "path polynomials" have rich structure and could provide interesting theoretical and physical applications.

Open Question 2. *What properties, in addition to symmetry, do classes of "path polynomials", with interesting theoretical and physical applications, have and what are the maximal polynomials for various values of input?*

We now turn to some analagous considerations for multidimensional recurrences and related multivariate power series.

Table 4.2. Optimal Polynomials

$U_i\backslash A$	$[B,\infty)\cap\left(\frac{1}{J(1)},\infty\right)$	$[B,\infty)\cap\left(\frac{1}{J(2)},\frac{1}{J(1)}\right]$	$[B,\infty)\cap\left(\frac{1}{J(3)},\frac{1}{J(2)}\right]$	$[B,\infty)\cap\left(\frac{1}{J(4)},\frac{1}{J(3)}\right]$	$[B,\infty)\cap\left(\frac{1}{J(5)},\frac{1}{J(4)}\right]$	$\cdots$
U_1	1	1	1	1	1	$\cdots$
U_2	A	A	A	A	A	$\cdots$
U_3	$\phi_1(1)$	$AB+A$	$AB+A$	$AB+A$	$AB+A$	$\cdots$
U_4	$\phi_2(1,1)$	$\phi_2(1,1)$	$\phi_2(1,1)$	$\phi_2(1,1)$	$\phi_2(1,1)$	$\cdots$
U_5	$\phi_3(1,1,1)$	$\phi_1(3)$	$\phi_2(1,2)$	$\phi_2(1,2)$	$\phi_2(1,2)$	$\cdots$
U_6	$\phi_4(1,1,1,1)$	$\phi_2(2,2)$	$\phi_2(2,2)$	$\phi_2(2,2)$	$\phi_2(2,2)$	$\cdots$
U_7	$\phi_5(1,1,1,1,1)$	$\phi_3(2,1,2)$	$\phi_1(5)$	$\phi_2(2,3)$	$\phi_2(2,3)$	$\cdots$
U_8	$\phi_6(1,1,1,1,1,1)$	$\phi_4(2,1,1,2)$	$\phi_2(3,3)$	$\phi_2(3,3)$	$\phi_2(3,3)$	$\cdots$
U_9	$\phi_7(1,1,1,1,1,1,1)$	$\phi_5(2,1,1,1,2)$	$\phi_3(3,1,3)$	$\phi_1(7)$	$\phi_2(3,4)$	$\cdots$
U_{10}	$\phi_8(1,1,1,1,1,1,1,1)$	$\phi_6(2,1,1,1,1,2)$	$\phi_4(3,1,1,3)$	$\phi_2(4,4)$	$\phi_2(4,4)$	$\cdots$
	$\vdots$	$\vdots$	$\vdots$	$\vdots$	$\vdots$	$\ddots$

4. Multivariable Power Series with Restricted Coefficients

In this section we consider an extension to the concepts for one dimensional recurrences to higher dimensions. For further discussion of the methods used, see [7].

Suppose that for $\mathbf{n} = (n_1, n_2, \ldots, n_\nu) \in (\mathbb{Z}^+)^\nu$, (analagous to (1.4)), we have

$$b_{\mathbf{n}} = \sum_{\substack{\mathbf{m}<\mathbf{n} \\ d(\mathbf{n},\mathbf{m})\leq k}} \alpha_{\mathbf{n},\mathbf{m}} b_{\mathbf{m}}, \tag{4.1}$$

with $b_{\mathbf{0}} = 1$, under some ordering on $(\mathbb{Z}^+)^\nu$, some distance function d and some double sequence $\{\alpha_{\mathbf{i},\mathbf{j}}\}$ satisfying

$$\alpha_{\mathbf{i},\mathbf{j}} \in [-1, 0], \tag{4.2}$$

for all $\mathbf{i},\mathbf{j} \in (\mathbb{Z}^+)^\nu$. Specifically, it will be convenient to consider $\mathbf{m} < \mathbf{n}$, if and only if $m_i \leq n_i$ for all i and $\mathbf{m} \neq \mathbf{n}$. Also we shall take d to be the usual "taxicab" metric, i.e. $d(\mathbf{n},\mathbf{m}) = \sum_{1\leq i\leq \nu} |n_i - m_i|$.

Note that, more generally, one might consider recurrences on any partially ordered set P (see [17]). In that case the Möbius function of P, i.e. the inverse of its Zeta function, would be an example of a recursive function in the incidence algebra $I(P)$ with negative coefficients (cf. [51], [43], [30], [28]). In fact, the inverse of any function $f \in I(P)$, with $f(x,x) = 1$ for all x, that takes values in $[0,1]$, would have coefficients as in (4.2).

We are interested, now, in the value of $\{U_{\mathbf{n}}(k)\}$ where, for fixed k,

$$U_{\mathbf{n}}(k) \stackrel{def}{=} \max\{|b_{\mathbf{n}}| : \{b_{\mathbf{i}}\}, \{\alpha_{\mathbf{i},\mathbf{j}}\} \text{ satisfy } (4.1) \text{ and } (4.2)\}.$$

Determination of $\{U_{\mathbf{n}}(k)\}$ for $\nu = 1$ and finite k has been studied in [8], [9], [10], and [16].

The following theorem was proven in [7] via path counting techniques.

Theorem 4.1. *Suppose* $v \geq 1$ *and* $k \geq 1$ *is odd or infinite, and define* $\{V_{\mathbf{i}}(k)\}$ *by* $V_{\mathbf{1}} = 1$ *and*

$$V_{\mathbf{i}}(k) = \sum_{\substack{\mathbf{j}<\mathbf{i} \\ d(\mathbf{i},\mathbf{j})\equiv 1 \mod 2 \\ d(\mathbf{i},\mathbf{j})\leq k}} V_{\mathbf{j}}(k). \tag{4.3}$$

Then, $U_{\mathbf{i}}(k) = V_{\mathbf{i}}(k)$ *for all* $\mathbf{i}$.

Proof. See [7]. □

Note that for k even, the corresponding result in Theorem 4.1 does not hold (see the counterexamples in [8] for the case $v = 1$).

As in Section 1., the result in Theorem 4.1 can be applied to obtain results for reciprocals and zero-free regions for multivariate power series.

In particular, for a fixed $I \subset \mathfrak{R}$, let $\mathcal{F}_I^{(v)}$ be the set of v-variate, I-power series defined by

$$\mathcal{F}_I^{(v)} = \left\{ f : f(\mathbf{z}) = \sum_{\mathbf{i}\in\mathbb{N}^v} a_{\mathbf{i}}\mathbf{z}^{\mathbf{i}},\ a_0 = 1, \text{ and } a_{\mathbf{i}} \in I \text{ for each } \mathbf{i} \in (\mathbb{Z}^+)^v \right\}.$$

The results are most easily illustrated via the two-dimensional case (i.e. $v = 2$). For convenience, we write, for instance, $f_{i,j}$ in place of $f_{(i,j)}$.

As in (1.1), the coefficients of the multiplicative inverse h of a series $f = \sum\sum f_{i,j}x^iy^j \in \mathcal{F}_{[0,1]}^{(2)}$ may be computed by equating coefficients in the expansion

$$\left(\sum\sum h_{i,j}x^iy^j\right)\left(\sum\sum f_{i,j}x^iy^j\right) = 1. \tag{4.4}$$

This results in $h_{0,0} = 1$, and for $(n,m) \neq (0,0)$,

$$h_{n,m} = -\sum_{\substack{i\leq n;\ j\leq m \\ (i,j)\neq(n,m)}} f_{n-i,m-j}h_{i,j}. \tag{4.5}$$

In particular, if $f_{i,j} = 0$ whenever $i + j > k$ (i.e. f is a polynomial in x and y with total degree less than or equal to k), then the coefficient $f_{n-i,m-j}$ in (4.5) is zero whenever $d((n,m),(i,j)) > k$ and we have a kth order recurrence as in (4.1):

$$h_{n,m} = \sum_{\substack{(i,j)<(n,m) \\ d((n,m),(i,j))\leq k}} (-f_{n-i,m-j})h_{i,j}. \tag{4.6}$$

Theorem 4.1 is, then, directly applicable and leads to the following corollary which extends Theorem 1.3.

Corollary 4.2. *Suppose that* $f \in \mathcal{F}_{[0,1]}^{(2)}$ *has total degree k, with k odd or infinite, and f and h satisfy (4.4). Then,*

$$|h_{(i,j)}| \leq V_{(i+1,j+1)}(k) \tag{4.7}$$

for all $(i,j) \in \mathbb{N}^2$ *where* $\{V_{\mathbf{i}}(k)\}$ *is as in (4.3).*

Table 4.3. Some Values of $V_{(i,j)}(\infty)$

i/j	1	2	3	4	5	6	7
1	1	1	1	2	3	5	8
2	1	2	4	8	15	28	51
3	1	4	10	23	50	104	210
4	2	8	23	60	144	328	718
5	3	15	50	144	378	931	2187
6	5	28	104	328	931	2458	6150
7	8	51	210	718	2187	6150	16296

Table 2 gives values of $\{V_{\mathbf{i}}(\infty)\}$ for some small $\mathbf{i}$.

It is not difficult to show, from (4.3), that $\{V_{(i,j)}(\infty)\}$ satisfies the simpler recurrence

$$V_{(n_1,n_2)}(\infty) = V_{(n_1-1,n_2)}(\infty) + V_{(n_1-2,n_2)}(\infty) \\ + V_{(n_1,n_2-1)}(\infty) + V_{(n_1,n_2-2)}(\infty) - V_{(n_1-2,n_2-2)}(\infty) \quad (4.8)$$

whenever $n_1, n_2 > 3$.

As before, bounds of the sort in Corollary 4.2 may be useful where generating functions or formal power series are utilized such as in enumerative combinatorics and in the study of stochastic processes. For further discussion of multi-variable generating functions and related multivariate linear recurrences c.f. [36], [37], [38], and the references therein.

As mentioned, Flatto *et al.* [23] and Solomyak [44] independently proved that if z is a root of a series in $\mathcal{F}^{(1)}_{[0,1]}$, then $|z| \geq r_0^{-1}$, where $r_0 = (1+\sqrt{5})/2) \approx 1.618034$. The following related result is an immediate consequence of Corollary 4.2 and (4.8) and Abel's Lemma (c.f. [42], [41], [33]).

Corollary 4.3. *(Zero-free region for elements of $\mathcal{F}^{(2)}_{[0,1]}$). Suppose $f \in \mathcal{F}^{(2)}_{[0,1]}$, and (x_1, x_2) is inside the bi-disc $\{|x_j| < r^{-1}, j = 1,2\}$ where $r \approx 2.69173951$ is the largest real root of $x^4 - 2x^3 - 2x^2 + 1 = 0$. Then $h(x_1,x_2) = L < \infty$, and hence $f(x_1,x_2) = 1/L \neq 0$.*

Note that the series consisting of all odd degree terms, i.e.

$$f(x,y) = 1 + x + y + x^3 + x^2y + xy^2 + y^3 + x^5 + \cdots,$$

has a zero at $(x,y) = (-r^{-1}, -r^{-1})$ and the bi-disk given in the corollary is optimal.

Similar reasoning leads to the following result for general v.

Corollary 4.4. *(Zero-free region for elements of $\mathcal{F}^{(v)}_{[0,1]}$) Suppose $f \in \mathcal{F}^{(v)}_{[0,1]}$, and $\mathbf{x} = (x_1, x_2, \ldots, x_k)$ is inside the polydisk $\{|x_j| < r_v^{-1}, 1 \leq j \leq v\}$, where r_v is the largest real root of the polynomial g_v, given by*

$$g_v(X) = X^{2^v} - vX^{2^v-1} - vX^{2^v-2} + v - 1.$$

Then, $f(\mathbf{x}) \neq 0$.

Table 4.4. Radii of optimal polydisks

ν	r_ν	r_ν^{-1}
1	1.618033989	0.6180339887
2	2.691739510	0.3715069739
3	3.791140848	0.2637728431
4	4.828427125	0.2071067812
5	5.854101966	0.1708203933
6	6.872983346	0.1454972244
7	7.887482194	0.1267831705
8	8.898979486	0.1123724357
9	9.908326913	0.1009252126

Note that $g_\nu(1) < 0$ and g_ν is of even degree and hence $r_\nu \in (1, \infty)$.

As in the case $\nu = 2$, the series consisting of all odd degree terms has a zero at $\mathbf{x} = (-r_\nu^{-1}, -r_\nu^{-1}, \ldots, -r_\nu^{-1})$ and hence, the polydisk given in the corollary is optimal. Table 4.4 gives the (numerically computed) radii of the optimal zero-free polydisks for small ν.

Corollary 4.4 then leads to the following (see [7]).

Theorem 4.5. *Suppose $\nu \geq 1$, and let r_ν^{-1} be the radius of the optimal zero-free ν-dimensional polydisk defined in the statement of Corollary 4.4. Then, we have $\nu < r_\nu < \nu + 1$ and*

$$\lim_{\nu \to \infty} \frac{r_\nu}{\nu + 1} = 1.$$

Of course, the preceding discussion only deals with the specific case of coefficients as in (4.2). Results for other intervals for higher dimensions would be interesting and would have immediate implications for power series.

The question of behavior of $\{U_{\mathbf{n}}(k)\}$ for even k could also be of interest. For results for $k = 2$ and $\nu = 1$, see [16] and [11]. Consideration of the case $k = 4$ and $\nu = 1$ is included in [10].

Work is currently underway, as well, to obtain results on generalizations to recurrences on graphs and partially ordered sets (see [17]).

We close this section with two open questions.

Open Question 3. *For $V < W$, Define $\{U_{\mathbf{i}}(\infty)(V,W)\}$ by*

$$U_{\mathbf{n}}(\infty)(V,W) \stackrel{def}{=} \max\{|b_{\mathbf{n}}| : \{b_{\mathbf{i}}\} \text{ satisfies (4.1) and } \{\alpha_{\mathbf{i},\mathbf{j}}\} \in [V,W]\}.$$

What is the value of $\{U_{\mathbf{n}}(\infty)(V,W)\}$, or more simply, what are bounds on the rate of growth of this sequence?

Open Question 4. *For k even, what is the value of $\{U_{\mathbf{n}}(k)\}$, or more simply, what are bounds on the rate of growth of this sequence?*

We now turn to consideration of inequalities for power series with monotone coefficients.

5. Power Series with Monotone Coefficients (An Explicit Kendall's Theorem)

Some of the original motivation for studying difference equations with constrained coefficients arose in a probability setting where the coefficients are tail probabilities and hence form a monotone sequence. In this section we briefly discuss some recent work in this direction.

Firstly, we consider reciprocals of power series whose coefficients are *monotone* and bounded by a *geometrically* decaying sequence. In particular, for fixed $A \geq 1$ and $0 < r < 1$, let the sets $\mathcal{F}_{A,r}$ and $\mathcal{F}^I_{A,r}$ be defined by

$$\mathcal{F}_{A,r} \stackrel{def}{=} \Big\{ Q : Q(z) = 1 + \sum_{k=1}^{\infty} q_k z^k, \{q_k\} \text{ is nonincreasing and } 0 \leq q_i \leq Ar^i \text{ for each } i \geq 1 \Big\}, \tag{5.1}$$

and

$$\mathcal{F}^I_{A,r} \stackrel{def}{=} \left\{ \Omega : \Omega(z) = 1 + \sum_{k=1}^{\infty} \omega_k z^k \text{ and } \Omega = \frac{1}{Q}, \text{ for some } Q \in \mathcal{F}_{A,r} \right\}.$$

Disregarding its probabilistic context, the well-known Kendall's Renewal Theorem (cf. [25], [24], [31], [32], [12], [13]) can essentially be restated as follows.

Theorem 5.1 ([25]). *Suppose $\Omega(z) = \sum_{i=0}^{\infty} \omega_i z^i \in \mathcal{F}^I_{A,r}$, for some $A > 0$ and $r < 1$. Then,*

$$|\omega_i| = O\left(\sigma^i\right) \tag{5.2}$$

for some $\sigma < 1$.

In [3], steps are taken towards obtaining an explicit form of Theorem 1.1.

Theorem 5.2. *Suppose $\Omega \in \mathcal{F}^I_{A,r}$, where A and r are constants satisfying*

(1) $1 \leq A \leq 1/(2r)$,

(2) $\left(A^2r^2 - 2A^2r + A\right)\left(A + \frac{1/2}{A+1/2}\right) + Ar^2 - 2Ar - A^2r^3 + A^2r^2 \geq 0$,

(3) $\left(-2A^2r + A\right)\left(A + \frac{1-Ar}{A+1-Ar}\right) - Ar + A^2r^2 \geq 0$, *and*

(4) $\frac{(1-Ar)\left(A^2-A+1-Ar\right)}{A+1-Ar} + A\left(A^2 - A + 1 - Ar\right) + A - 1 - A^2r + Ar - A^2 \geq 0$.

If $\Omega(z) = \sum_{i=0}^{\infty} \omega_i z^i$, then for $n \geq 1$,

$$|\omega_n| \leq C_{A,r} s^n_{A,r},$$

where $\delta \stackrel{def}{=} \sqrt{A^2 + 4(1 - Ar)}$, $s_1 \stackrel{def}{=} \frac{\delta + A}{2}$,

$$s_{A,r} \stackrel{def}{=} rs_1 < 1, \quad \textit{and} \quad C_{A,r} \stackrel{def}{=} A/s_1 \leq 1.$$

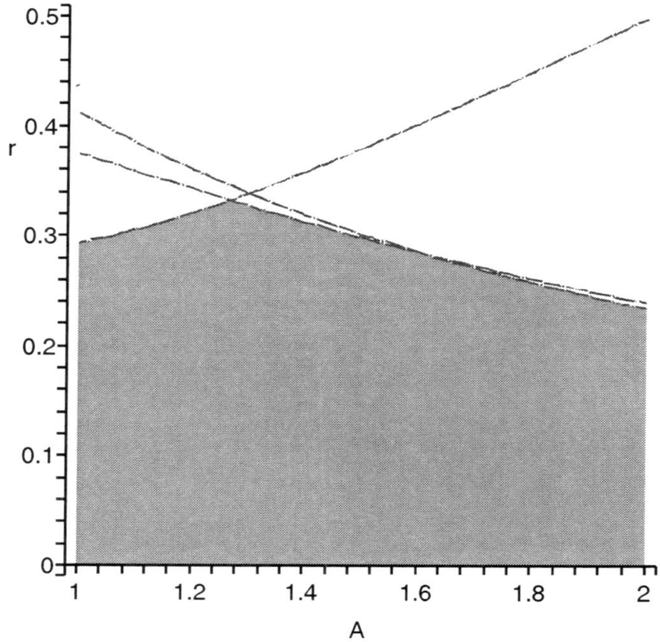

Figure 4.1. Some pairs (A, r) satisfying the assumptions of Theorem 5.2

Proof. See [3]. □

Note that if (A, r) satisfies $Ar < 1$, then $s_{A,r} < 1$ and (as suggested by Theorem 5.1) the coefficients of the reciprocal series decay at an exponential rate.

Some pairs (A, r) satisfying the assumptions of Theorem 5.2 are given in Figure 4.1.

For pairs (A, r) satisfying Assumptions (1)–(4), Theorem 5.2 gives the optimal value of σ in (5.2) [Kendall's Theorem] which applies for all $\Omega \in \mathcal{F}^{I}_{A,r}$.

As before, Theorem 5.2 has immediate implications for lower bounds for the modulus of the smallest zero of a power series $Q \in \mathcal{F}_{A,r}$. Since the result places a lower bound on the radius of convergence of $\frac{1}{Q}$, it also places a lower bound on zeroes of Q. Indeed, we have the following corollary.

Corollary 5.3. *Suppose z_0 is a root of a power series $Q \in \mathcal{F}_{A,r}$, where $Q(z) = 1 + q_1 z + q_2 z^2 + \cdots$, and A and r satisfy the hypotheses of Theorem 5.2, then*

$$|z_0| \geq s_{A,r}^{-1}.$$

Proof. See [3]. □

In fact, the series $Q \in \mathcal{F}_{A,r}$ given by

$$Q(z) = 1 + Arz + Ar^3z^2 + Ar^3z^3 + Ar^5z^4 + Ar^5z^5 + Ar^7z^6 + \cdots \qquad (5.3)$$

has a zero at $z = -s_{A,r}^{-1}$, and Corollary 5.3 is optimal. The series in (5.3) also serves to show that the decay rate $s_{A,r}$ of Theorem 5.2 is optimal in that context, as well.

Consideration of the remaining pairs (A, r) not covered by Theorem 5.2, as well as applications to Markov chains and renewal convergence, are among potential questions for further investigation.

Consideration of other coefficient properties including convexity and quasi-monotonicity is currently under way. For some results involving recurrences with monotone coefficients c.f. [4] and [6].

As in Section 4., we close this section with two open questions.

Open Question 5. (Explicit Kendall's Theorem) *For pairs (A, r) not satisfying Assumptions (1)–(4), of Theorem 5.2, what is the optimal value of σ in (5.2) [Kendall's Theorem]?*

Open Question 6. *Define*

$$\mathcal{F}_{A,convex} \stackrel{def}{=} \left\{ Q : Q(z) = 1 + \sum_{k=1}^{\infty} q_k z^k, \{q_k\} \text{ is convex and } 0 \le q_i \le A \right\}.$$

For $Q \in \mathcal{F}_{A,convex}$, what is the maximal absolute value of the nth coefficient of $1/Q$?

Bibliography

[1] F. Beaucoup, P. Borwein, D.W. Boyd and C. Pinner, Power series with restricted coefficients and a root on a given ray, *Mathematics of Computation*, **222** (1998), 715–736.

[2] F. Beaucoup, P. Borwein, D.W. Boyd and C. Pinner, Multiple roots of [-1,1] power series, *Journal of the London Mathematical Society*, **57** (1998), 135–147.

[3] K.S. Berenhaut, E.E. Allen and S.J. Fraser, Bounds on coefficients of reciprocals of formal power series with rapidly decreasing coefficients, in press, *Discrete Dynamics in Nature and Society*, (2006).

[4] K.S. Berenhaut and D. Bandyopadhyay, Monotone convex sequences and Cholesky decomposition of symmetric Toeplitz matrices, *Linear Algebra and its Applications*, **403** (2005), 75–85.

[5] K.S. Berenhaut, Y.W. Fan and Z. Abernathy, On optimization for linear recurrence relations with unbounded order, preprint (2005).

[6] K.S. Berenhaut and P.T. Fletcher, On inverses of triangular matrices with monotone entries, *Journal of Inequalities in Pure and Applied Mathematics*, **6**(4) (2005), Art. 63.

[7] K.S. Berenhaut and J.D. Foley, Explicit bounds for multi-dimensional linear recurrences with restricted coefficients, in press, *Journal of Mathematical Analysis and Applications,* **322**(2) (2006), 1159–1167.

[8] K.S. Berenhaut and J.D. Foley, Bounds for linear recurrences with coefficients in the negative unit interval and an application to $(0, 1)$ matrices, submitted, *Discrete Mathematics,* (2005).

[9] K.S. Berenhaut and J.D. Foley, Applications of recurrence bounds to networks and paths, in press, *International Journal of Applied Mathematics*, (2006).

[10] K.S. Berenhaut, B.G. Gibson, J.H. Newman and J.F. Andreson, Bounds for fourth order (0,1) difference equations, submitted, *Computers and Mathematics with Applications,* (2005).

[11] K.S. Berenhaut and E.G. Goedhart, Second-order linear recurrences with restricted coefficients and the constant $(1/3)^{1/3}$, *Mathematical Inequalities & Applications,* **9** (2006), 445–452.

[12] K.S. Berenhaut and R.B. Lund, Geometric renewal convergence rates from hazard rates, *Journal of Applied Probability*, **38** (2001), 180–194.

[13] K.S. Berenhaut and R.B. Lund, Renewal convergence rates for dhr and nwu lifetimes, *Probability in the Engineering and Informational Sciences*, **16**(1) (2002), 67–84.

[14] K.S. Berenhaut and R.B. Lund, Bounds for linear recurrences with restricted coefficients, *Journal of Inequalities in Pure and Applied Mathematics*, **4**(2) (2003), Article 26, 15 pp.

[15] K.S. Berenhaut and D.C. Morton, Second-order bounds for linear recurrences with negative coefficients, in press, *Journal of Computational and Applied Mathematics*, **186**(2) (2006), 504–522.

[16] K.S. Berenhaut, D.C. Morton and Y.W. Fan, Bounds for second-order recurrences in terms of maximal products over integer partitions, *Proceedings of the Eleventh International Conference on Fibonacci Numbers*, in press, (2005).

[17] K.S. Berenhaut, A.B. O'Keefe and F. Saidak, Bounds for recurrences on ranked posets, preprint, (2006).

[18] P. Borwein, and C.G. Pinner, Polynomials with 0,+1,-1 coefficients and a root close to a given point, *Canadian Journal of Mathematics*, **49** (1997), 887–915.

[19] P. Borwein and T. Erdélyi, On the zeros of polynomials with restricted coefficients, *Illinois Journal of Mathematics*, **46** (1997), 667–675.

[20] M. Embree and L.N. Trefethen, Growth and decay of random Fibonacci sequences, *The Royal Society of London Proceedings, Series A, Mathematical, Physical and Engineering Sciences*, **455** (1999), 2471–2485.

[21] Y.W. Fan and K.S. Berenhaut, A bound for linear recurrence relations with unbounded order, *Computers and Mathematics with Applications,* **50**(3-4) (2005), 509–518.

[22] W. Feller, *An Introduction to Probability Theory and its Applications, Volume I*, 3rd Edition, John Wiley and Sons, New York (1968).

[23] L. Flatto, J.C. Lagarias and B. Poonen, Lap numbers and periodic points of the beta transformation, *Ergodic Theory and Dynamical Systems*, (2) **14** (1994), 237–266.

[24] C.R. Heathcote, Complete exponential convergence and related topics, *Journal of Applied Probability*, **4** (1967), 1–40.

[25] D.G. Kendall, Unitary dilations of Markov transition operators and the corresponding integral representations for transition probability matrices, in: *Probability and Statistics*, Edited by U. Grenander, 139-161, Wiley, New York, (1959).

[26] M. Kijima, *Markov Processes for Stochastic Modeling*, Chapman and Hall, London (1997).

[27] R.K. Kittappa, A representation of the solution of the nth order linear difference equations with variable coefficients, *Linear Algebra and Applications*, **193** (1993), 211–222.

[28] D.N. Kozlov, Maximizing the Möbius function of a poset and the sum of the Betti numbers of the order complex, *Combinatorica,* **19**(4) (1999), 533–548.

[29] R.K. Mallik, On the solution of a linear homogeneous difference equation with varying coefficients, *SIAM Journal on Mathematical Analysis*, **31** (2000), 375–385.

[30] E.E. Marenich, Bound for values of the Möbius function, *Mat. Zametki,* **44**(4) (1988), 469–487.

[31] S.P. Meyn and R.L. Tweedie, *Markov Chains and Stochastic Stability*, Springer-Verlag, London, (1993).

[32] S.P. Meyn and R.L. Tweedie, Computable bounds for convergence rates of Markov chains, *Annals of Applied Probability*, **4** (1994), 981–1011.

[33] R. Narasimhan, *Several Complex Variables. Chicago Lectures in Mathematics*. The University of Chicago Press, Chicago, Ill.-London, (1971).

[34] A.M. Odlyzko, Asymptotic enumeration methods, in: *Handbook of Combinatorics* (R. Graham, M. Groetschel, and L. Lovasz, Eds.), Volume II, Elsevier, Amsterdam, (1995), 1063–1229.

[35] A. Odlyzko and B. Poonen, Zeros of polynomials with 0,1 coefficients, *L'Enseignement Mathematique*, **39** (1993), 317–348.

[36] R. Pemantle and M.C. Wilson, Asymptotics of multivariate sequences I: smooth points of the singular variety, *Journal of Combinatorial Theory A,* **97** (2002), 129–161.

[37] R. Pemantle and M.C. Wilson, Asymptotics of multivariate sequences II: multiple points of the singular variety, *Combinatorics, Probability and Computing,* **13**(4-5) (2004), 735–761.

[38] M. Petkovšek and M. Bousquet-Mélou, Linear recurrences with constant coefficients: the multivariate case, *Discrete Mathematics*, **225**(1-3) (2000), 51–75.

[39] C.G. Pinner, Double roots of [-1,1] power series and related matters, *Mathematics of Computation*, **68** (1999), 1149–1178.

[40] J. Popenda, One expression for the solutions of second order difference equations, *Proceedings of the American Mathematical Society*, **100** (1987), 87–93.

[41] R. Michael Range, Complex analysis: a brief tour into higher dimensions, *Amer. Math. Monthly,* **110**(2) (2003), 89–108.

[42] R. Michael Range, *Holomorphic functions and integral representations in several complex variables*. *Graduate Texts in Mathematics*, **108**, Springer-Verlag, New York, (1986).

[43] B.E. Sagan, Yeong Nan Yeh and G.M. Ziegler, Maximizing Möbius functions on subsets of Boolean algebras, *Discrete Math.*, **126**(1-3) (1994), 293–311.

[44] B. Solomyak, Conjugates of beta-numbers and the zero-free domain for a class of analytic functions, *Proceedings of the London Mathematical Society*, **68** (1994), 477–498.

[45] D. Viswanath, *Lyapunov exponents from random Fibonacci sequences to the Lorenz equations*, Ph.D. Thesis, Department of Computer Science, Cornell University, (1998).

[46] D. Viswanath, Random Fibonacci sequences and the number 1.13198824..., *Mathematics of Computation*, **69** (2000), 1131–1155.

[47] D. Viswanath and L.N. Trefethen, Condition numbers of random triangular matrices, *SIAM Journal on Matrix Analysis and Applications*, **19** (1998), 564–581.

[48] H.S. Wilf, *Generatingfunctionology*, Second Edition, Academic Press, Boston, (1994).

[49] T.G. Wright and L.N. Trefethen, Computing Lyapunov constants for random recurrences with smooth coefficients, *Journal of Computational and Applied Mathematics*, **132**(2) (2001), 331–340.

[50] O. Yamamoto, On some bounds for zeros of norm-bounded polynomials, *Journal Symbolic Computation*, **18**(5) (1994), 403–427.

[51] G.M. Ziegler, Posets with maximal Möbius function, *J. Combin. Theory Ser. A*, **56**(2) (1991), 203–222.

In: Advances in Inequalities for Series
Editors: S. S. Dragomir and A. Sofo, pp. 89–120
ISBN 978-1-60021-920-7

Chapter 5

Inequalities in Series and Summation in Hypernumbers

Mark Burgin
Department of Mathematics
University of California, Los Angeles
405 Hilgard Ave., Los Angeles, CA 90095
E-mail address: mburgin@math.ucla.edu

Abstract

Inequalities play an important role for series. Here we consider inequalities in the context of hypernumbers. The theory of hypernumbers is a novel approach in functional analysis aimed at the development of mathematically correct techniques, which allow operations with integrals and series divergent in the classical sense. In the second section of this work, basic notions and constructions of the theory of hypernumbers are introduced. This forms a base for the system development of operations on series in hypernumbers in the third section. One of the most important operations on series is summation. In this paper, two types of summation are considered: topological and analytical. This allows one to find the analytical sum of an arbitrary series of real or complex numbers and to determine when topological and analytical sums coincide. In the fourth section, local and global inequalities for series are studied. It is proved that if a series does not converge in the classical sense, while the absolute value of its analytical sum is bounded, then for any hypernumber α there is a quotient series of a permutation of a quotient series of the initial series that converges to α. In the fifth section, relations between inequalities and operations on series are studied.

1991 Mathematics Subject Classification: 40.

Key words and phrases: Series, Summation, Inequality, Hypernumber, Infinite hypernumber, Local inequality, Global inequality.

1. Introduction

We consider inequalities for series in a new context when it is possible to find a sum of any series of real or complex numbers. This becomes attainable due to utilization of hypernumbers. Summation of series is traditionally defined by utilizing limits of sequences.

This reduces many results for series to problems with sequences. For instance, many series do not have a sum because, as we know from any course of calculus, many sequences do not have limits. However, the goal to overcome limitations and restrictions with respect to operation with mathematical structures has always been one of the main stimuli for the development of mathematics. For instance, when people knew only natural numbers, they were not able to perform subtraction for any pair of numbers. To overcome this limitation, zero and negative numbers were introduced and the number universe was extended from natural numbers to integer numbers. Introduction of rational numbers provided a possibility to divide a natural number by any natural number. Introduction of complex numbers aimed at solving a simple equation $x^2 = -1$ brought a possibility to solve arbitrary algebraic equations.

Thus, we may ask a question how it is possible to extend the existing number universe to get a limit to any sequence. A natural way to do this is to extend the set of all real numbers to the set of all real hypernumbers, the set of all complex numbers to the set of all complex hypernumbers, and sets of numerical functions to sets of extrafunctions and hyperdistributions (cf., for example, [4, 5]).

Introduction of hypernumbers was rooted in problems of contemporary physics. Many mathematical models used in modern field theories (e.g., gauge theories) result in divergence for analytically calculated properties of physical systems. Different methods of divergence elimination (regularization), i.e., of elimination of infinity, have been elaborated. Nevertheless, the majority of them are not well grounded mathematically because they utilize operations that have neither mathematical nor physical meaning. Moreover, there are such models in physics that contain infinities that cannot be eliminated by existing mathematical and physical theories [12]. There was a hope that string theory, which is an outgrowth of quantum field theory, would eliminate all divergencies [10]. Unfortunately, it did not happen. The strings under scrutiny are almost exclusively those defined in 10 space-time dimensions. Recently these strings were discovered to be membranes in disguise, defined in 11 dimensions. These theories have the problem that membranes are infinite even in perturbation theory. Infinity of membranes corresponds to infinities appearing only nonperturbatively in 10 dimensional strings, and thus previously unrecognized [20]. Thus, physicists were led by necessity to work with infinite quantities in theory and to come through them to finite measurable values. The first step in this direction were distributions [19], which formalized ideas of physicists about functions with infinite values. Hypernumbers and extrafunctions continue mathematical development in this direction.

Hypernumbers proved themselves useful in many aspects. First, they provided for an essential extension of the power of differential calculus [4]. In the extended calculus of hypernumbers and extrafunctions, it is possible, for example, to differentiate any real function. Second, hypernumbers form a base for the development of integral calculus [5], giving a natural formal setting for path integration and essentially extending the scope of integrable functions. Third, differential equations are important and powerful tools for many disciplines: physics, biology chemistry, economics and so on. Hypernumbers and, based on them, extrafunctions provided means to solve many types of differential equations that are unsolvable in the conventional setting (even with distributions) [8].

It is also necessary to remark that although real hypernumbers look similar to hyperreal numbers of nonstandard analysis [16], they are different and have rather dissimilar properties. The cause for this is that hyperreal numbers from nonstandard analysis are built based on set theoretical constructions, while real hypernumbers are built based on topological constructions (cf., for example, [4]).

In addition to a new theory, this paper gives a more rigorous exposition of series than standard textbooks on calculus. Thus, it might be useful for the development of teaching calculus in colleges and universities.

We consider two types of inequalities for series: local and global. Local inequalities show order relation between separate elements of series. Global inequalities reflect order relation between sums of series.

For simplicity, we consider here only real hypernumbers, sequences, and series. It is possible to develop a similar theory for the complex case.

Denotations:

- $\mathbb{N}$ is the set of all natural numbers;
- ω is the sequence of all natural numbers;
- $\emptyset$ is the empty set;
- $\mathbb{R}$ is the set of all real numbers;
- $\mathbb{Q}$ is the set of all rational numbers;
- $\mathbb{R}^+$ is the set of all non-negative real numbers;
- $\mathbb{R}^{++}$ is the set of all positive real numbers;
- If a is a real number, then $|a|$ or $||a||$ denotes its absolute value or modulus;
- $\mathbb{R}^\omega$ is the set of all sequences of real numbers;
- $\mathbb{R}_\omega$ is the set of all real hypernumbers;
- $\mathbb{R}_\omega^+$ is the set of all real hypernumbers that are larger than or equal to zero;
- $\mathbf{Ser}_\mathbb{R}$ is the space of all real number series.

2. Spaces of Hypernumbers

Here we give only constructions and results from the theory of hypernumbers utilized in our study of inequalities for series. It is possible to find a more detailed theory, including proofs of results given in this section in [4, 5].

Let $\mathbb{R}^\omega = \{(a_i)_{i\in\omega}; a_i \in \mathbb{R}\}$ be the set of all sequences of real numbers.

Definition 1. *For arbitrary sequences $a = (a_i)_{i\in\omega}$, $b = (b_i)_{i\in\omega} \in R^\omega$: $a \sim b$ means that* $\lim_{i\to\infty} |a_i - b_i| = 0$.

Lemma 2.1. *The relation $\sim$ is an equivalence.*

Definition 2. *Classes of the equivalence $\sim$ are called real hypernumbers and their set is denoted by $\mathbb{R}_\omega$.*

Any sequence $a = (a_i)_{i\in\omega}$ determines a hypernumber $\alpha = \mathrm{Hn}(a_i)_{i\in\omega}$. It is possible to build in the same way complex hypernumbers. They are sets of equivalent sequences of complex numbers.

Lemma 2.2. *If $a_i = b_i$ for almost all i, then $a = (a_i)_{i\in\omega} \sim b = (b_i)_{i\in\omega}$.*

Relations on $\mathbb{R}$ induce corresponding relations on $\mathbb{R}^\omega$.

Definition 3. *If $a, b \in \mathbb{R}^\omega$, then:*

a) $a \leq b$ means that $\exists n\ \forall i \geq n\ (a_i \leq b_i)$;

b) $a < b$ means that $\exists n\ \forall i \geq n\ (a_i < b_i)$.

These relations are a partial order and a strict partial order, respectively. They induce similar relations on $\mathbb{R}_\omega$:

Definition 4. *If α, $\beta \in \mathbb{R}_\omega$, then*

a) $\alpha \leq \beta$ means that $\exists a \in \alpha \quad \exists b \in \beta\ (a \leq b)$;

b) $\alpha < \beta$ means that $(\exists a \in \alpha \quad \exists b \in \beta\ (a < b))$ and $\alpha \neq \beta$.

Some operations in $\mathbb{R}$ induce corresponding operations in $\mathbb{R}^\omega$.
Let $a = (a_i)_{i\in\omega}$ and $b = (b_i)_{i\in\omega}$ be elements from $\mathbb{R}^\omega$ and $d \in \mathbb{R}$.

Definition 5.

a) Operation of addition in $\mathbb{R}^\omega : a + b = (c_i)_{i\in\omega}$, where $c_i = a_i + b_i$ for all $i \in \omega$;

b) Operation of subtraction in $\mathbb{R}^\omega : a - b = (c_i)_{i\in\omega}$, where $c_i = a_i - b_i$ for all $i \in \omega$;

c) Operation of multiplication by a real number: $d \cdot b = (c_i)_{i\in\omega}$, where $c_i = d \cdot b_i$ for all $i \in \omega$.

d) Operation of multiplication in $\mathbb{R}^\omega : a \cdot b = (c_i)_{i\in\omega}$, where $c_i = a_i \cdot b_i$ for all $i \in \omega$.

By the definition of addition and multiplication in the set $\mathbb{R}^\omega$, all laws of operations with real numbers (commutativity of addition, associativity of addition, commutativity of multiplication, associativity of multiplication, and distributivity) are valid for corresponding operations with sequences of real numbers.

Some operations in $\mathbb{R}^\omega$ induce similar operations in $\mathbb{R}_\omega$. As a result [4, 5], the set $\mathbb{R}_\omega$of all real hypernumbers is an ordered linear infinite dimensional space over the field $\mathbb{R}$ of real numbers, in which the binary operations *max* and *min* are also defined.

In contrast to addition and subtraction, multiplication of sequences in $\mathbb{R}^\omega$ does not determine multiplication in $\mathbb{R}_\omega$. For example, $0 = \mathrm{Hn}(z_i = 0)_{i\in\omega} = \mathrm{Hn}(1/i)_{i\in\omega}$. If we multiply both sequences by the sequence $(i)_{i\in\omega}$, we come to two different and not equivalent sequences. Namely, $\mathrm{Hn}(0 \cdot i)_{i\in\omega} = 0 \neq \mathrm{Hn}((1/i) \cdot (i))_{i\in\omega} = 1$.

Definition 6.

a) *A sequence* $(b_i)_{i \in H}$ *is called stable if there is a real number b such that* $b_i = b$ *for all* $i \in \omega$.

b) *A hypernumber* α *is called stable if it is defined by a stable sequence.*

We identify elements from $\mathbb{R}$ with stable hypernumbers from $\mathbb{R}_\omega$. For such a hypernumber, we will assume $\alpha = b$.

Proposition 2.3. *The operation of multiplication by stable sequences from* $\mathbb{R}^\omega$ *induces operation of multiplication by elements of* $\mathbb{R}$ *in* $\mathbb{R}_\omega$.

By the definition of addition and multiplication by stable sequences in the set $\mathbb{R}_\omega$, all laws of operations with real numbers (commutativity and associativity of addition, commutativity and associativity of multiplication by stable sequences, and distributivity with respect to multiplication by stable sequences) are valid for corresponding operations with real hypernumbers.

Real numbers contain different subclasses: rational, irrational, transcedental, integer, etc. In the universe of hypernumbers $\mathbb{R}_\omega$, there are even more subclasses. Thus, it is possible to introduce three basic types of hypernumbers: stable, infinite, and oscillating hypernumbers. Here are some examples.

Example 1. *An infinite increasing hypernumber:* $\alpha = \text{Hn}(a_i)_{i \in \omega}$, *where* $a_i = i$, $i = 1, 2, \ldots$.

Example 2. *An infinite increasing hypernumber:* $\beta = \text{Hn}(b_i)_{i \in \omega}$, *where* $b_i = 2^i$, $i = 1, 2, \ldots$.

Example 3. *A finite (bounded) oscillating hypernumber:* $\gamma = \text{Hn}(a_i)_{i \in \omega}$, *where* $a_i = (-1)^i$, $i = 1, 2, \ldots$.

Example 4. *An infinite oscillating hypernumber:* $\delta = \text{Hn}(a_i)_{i \in \omega}$, *where* $a_i = (-1)^i \cdot i$, $i = 1, 2, 3, \ldots$.

Example 5. *An infinite decreasing hypernumber:* $\nu = \text{Hn}(a_i)_{i \in \omega}$, *where* $a_i = -5i$, $i = 1, 2, 3, \ldots$.

Example 6. *An infinite oscillating hypernumber:* $\theta = \text{Hn}(a_i)_{i \in \omega}$, *where* $a_i = 2^i + (-1)^i \cdot i$, $i = 1, 2, 3, \ldots$.

Definition 7. *A real hypernumber* α *is called finite or bounded if there is a sequence* $(a_i)_{i \in \omega}$ *such that* $\alpha = \text{Hn}(a_i)_{i \in \omega}$ *and for some positive real number b, we have* $|b_i| < b$ *for almost all* $i \in \omega$.

Lemma 2.4. *The following conditions are equivalent:*

(1) α *is a finite real hypernumber;*

(2) *there is a sequence* $(a_i)_{i\in\omega}$ *such that* $\alpha = \text{Hn}(a_i)_{i\in\omega}$ *and for some real number* b, $|b_i| < b$ *for all* $i \in \omega$;

(3) *for any sequence* $(a_i)_{i\in\omega}$ *such that* $\alpha = \text{Hn}(a_i)_{i\in\omega}$ *there is a real number* b *such that* $|b_i| < b$ *for almost all* $i \in \omega$;

(4) *for any sequence* $(a_i)_{i\in\omega}$ *such that* $\alpha = \text{Hn}(a_i)_{i\in\omega}$ *there is a real number* b *such that* $|b_i| < b$ *for all* $i \in \omega$.

Lemma 2.5. *Any stable hypernumber* $\alpha = \text{Hn}(a_i)_{i\in\omega}$ *is finite.*

As stable real hypernumbers that are corresponded to different real numbers cannot be equal, we have the following result.

Lemma 2.6. *There is a one-to-one correspondence* f_R *between* $\mathbb{R}$ *and the subset* $\text{St}\mathbb{R}_\omega$ *of all stable hypernumbers from* $\mathbb{R}_\omega$ *defined by the formula* $f_R(a) = \text{Hn}(a_i)_{i\in\omega}$ *with all* $a_i = a$.

Definition 8. *A real hypernumber* $\alpha = \text{Hn}(a_i)_{i\in\omega}$ *is called oscillating if there is such* $k \in \mathbb{R}^{++}$ *that there are two infinite sequences of natural numbers* m_i *and* n_i *with* $i = 1, 2, \ldots$ *such that* $a_{m_i} - a_{n_i} > k$ *for all* $i = 1, 2, \ldots$

Oscillating hypernumbers may be bounded or finite (cf. Example 3) and unbounded or infinite (cf. Examples 4 and 6).

Definition 9. *A real hypernumber* α *is called:*

1) *infinite increasing if* $\alpha = \text{Hn}(a_i)_{i\in\omega}$ *for a sequence* $(a_i)_{i\in\omega}$ *such that*

$$\exists j \in \omega \quad \forall i > j(a_{i+1} - a_i > 0)) \quad \textit{and} \quad (\forall p \in \mathbb{R}\exists i \in \omega(a_i > p));$$

2) *infinite decreasing if* $\alpha = \text{Hn}(a_i)_{i\in\omega}$ *for a sequence* $(a_i)_{i\in\omega}$ *such that*

$$\exists j \in \omega \quad \forall i > j(a_{i+1} - a_i > 0)) \quad \textit{and} \quad (\forall p \in \mathbb{R}\exists i \in \omega(a_i < p));$$

3) *infinite expanding if* $\alpha = \text{Hn}(a_i)_{i\in\omega}$ *and there are subsequences* $(b_i)_{i\in\omega}$ *and* $(c_i)_{i\in\omega}$ *of the sequence* $(a_i)_{i\in\omega}$ *such that* $\text{Hn}(b_i)_{i\in\omega}$ *is an infinite increasing and* $\text{Hn}(c_i)_{i\in\omega}$ *is an infinite decreasing hypernumber;*

4) *strictly infinite expanding if it is infinite expanding and* $(a_i)_{i\in\omega} = (b_i)_{i\in\omega} \cup (c_i)_{i\in\omega}$.

Proposition 2.7. *Any finite real hypernumber is either a real number or an oscillating real hypernumber.*

Proposition 2.8. *Any infinite real hypernumber is either an infinite increasing number or an infinite decreasing number or an oscillating real hypernumber.*

Proposition 2.9. *If $\alpha = \mathrm{Hn}(a_i)_{i\in\omega}$ is a real number (an infinite increasing or infinite decreasing hypernumber), $\beta = \mathrm{Hn}(b_i)_{i\in\omega}$, and for almost all $i \in \omega$, the element b_i is between a_{i+1} and a_i, then β is a real number (an infinite increasing or infinite decreasing hypernumber).*

For oscillating hypernumbers, this is not always true as is demonstrated in the following example.

Example 7. *The condition of the Proposition 2.9 is satisfied for the pair that consists of the finite oscillating hypernumber $\alpha = \mathrm{Hn}(a_i)_{i\in\omega}$, $a_i = (-1)^i$, $i = 1,2,\ldots$ and for the real number $\beta = \mathrm{Hn}(b_i)_{i\in\omega}$, $b_i = 0$, $i = 1,2,\ldots$ However, β is not an oscillating hypernumber.*

The set of all rational numbers is dense in the space of all real numbers. This property gives us the following result.

Lemma 2.10. *Any class of equivalent sequences contains a sequence, all elements of which are rational numbers.*

This means that to generate all real hypernumbers, we need only sequences of rational numbers. In other words, all sequences of rational numbers give us all real hypernumbers in the way that all Cauchy sequences of rational numbers give us all real numbers.

Hypernumbers have specific invariants. One of the most important of them is the spectrum of a hypernumber.

Definition 10. *The spectrum* Spec a *of a sequence $a = (a_i)_{i\in\omega}$ is the set*

$$\{r \in \mathbb{R}; r = \lim b_i \text{ for some subsequence } (b_i)_{i\in\omega} \text{ of } a\}.$$

Definition 11. *The spectrum* Spec α *of a real hypernumber $\alpha = \mathrm{Hn}(a_i)_{i\in\omega}$ is equal to the set* Spec a *where $a = (a_i)_{i\in\omega}$.*

Example 8. *If $a = (a_i)_{i\in\omega}$ and $\alpha = \mathrm{Hn}(a_i)_{i\in\omega}$, where $a_i = i$, $i = 1,2,\ldots$, then* $\mathrm{Spec}\, a = \mathrm{Spec}\, \alpha = \emptyset$.

Example 9. *If $a = (a_i)_{i\in\omega}$ and $\alpha = \mathrm{Hn}(a_i)_{i\in\omega}$, where $a_i = (-1)^i$, $i = 1,2,\ldots$, then* $\mathrm{Spec}\, a = \mathrm{Spec}\, \alpha = \{1,-1\}$.

Example 10. *If $a = (a_i)_{i\in\omega}$ and $\alpha = \mathrm{Hn}(a_i)_{i\in\omega}$, where $a_i = (-1)\cdot i$, $i = 1,2,\ldots$, then* $\mathrm{Spec}\, a = \mathrm{Spec}\, \alpha = \emptyset$.

Example 11. *If $a = (a_i)_{i\in\omega}$ and $\alpha = \mathrm{Hn}(a_i)_{i\in\omega}$, where $a_i = 1/i$, $i = 1,2,\ldots$, then* $\mathrm{Spec}\, a = \mathrm{Spec}\, \alpha = \{0\}$.

Example 12. *If $a = (a_i)_{i\in\omega}$ and $\alpha = \mathrm{Hn}(a_i)_{i\in\omega}$, where $a_i = \sin(i/2)\pi$, $i = 1,2,\ldots$, then* $\mathrm{Spec}\, a = \{1,0,-1\}$.

Lemma 2.11. *If* $a = (a_i)_{i\in\omega} \sim b = (b_i)_{i\in\omega}$, *then* $\mathrm{Spec}\, a = \mathrm{Spec}\, b$.

Corollary 2.12. *For any hypernumber* α, *the set* $\mathrm{Spec}\,\alpha$ *does not depend on the choice of a sequence* $(a_i)_{i\in\omega}$ *from* α.

Proposition 2.13. *For any hypernumber* α, $\mathrm{Spec}\,\alpha$ *is a closed set.*

The spectrum of a hypernumber is a characteristic property, which allows one to discern different types of hypernumbers.

Proposition 2.14. $\alpha = \mathrm{Hn}(a_i)_{i\in\omega}$ *is a bounded oscillating real hypernumber if and only if* $\mathrm{Spec}\,\alpha$ *contains more than one element and the sequence* $(a_i)_{i\in\omega}$ *is bounded.*

Corollary 2.15. $\alpha = \mathrm{Hn}(a_i)_{i\in\omega}$ *is a stable real hypernumber if and only if* $\mathrm{Spec}\,\alpha$ *contains one element and the sequence* $(a_i)_{i\in\omega}$ *is bounded.*

Proposition 2.16. *Two stable real hypernumbers* α *and* β *are equal if and only if* $\mathrm{Spec}\,\alpha = \mathrm{Spec}\,\beta$.

For other kinds of real hypernumbers, this is not true. For example, the real hypernumbers $\alpha = \mathrm{Hn}(a_i)_{i\in\omega}$ with $a_i = (-1)^i$, $i = 1,2,\ldots$, and $\beta = \mathrm{Hn}(b_i)_{i\in\omega}$ with $b_i = (-1)^{i+1}$, $i = 1,2,\ldots$, have the same spectrum $\mathrm{Spec}\,\alpha = \mathrm{Spec}\,\beta = \{1,-1\}$.

Proposition 2.17. *The spectrum* $\mathrm{Spec}\,\alpha$ *is empty if and only if it is either an infinite increasing or infinite decreasing or strictly infinite expanding hypernumber.*

3. Operations on Series

In calculus, as a rule (cf., for example, [17] or [22]), a distinction between a series and its sum is not made. For instance, in all textbooks, it is usually written

$$\sum_{i=1}^{\infty} a_i = S \quad \text{where} \quad S = \lim_{n\to\infty} S_n \quad \text{where} \quad S_n = a_1 + a_2 + \cdots + a_n.$$

Thus, a series and its sum are treated as the same object. However, there is a necessity to make a distinction between a series and its sum because there are different definitions of a sum of a series. Some authors, however, develop a more rigorous approach. For example, Diedonne [11, Ch. 5] defines a series as a pair of sequences $(\{a_i; i = 1,2,\ldots\}, \{s_i; i = 1,2,\ldots\})$ where $s_i = a_1 + a_2 + \cdots + a_i$ for all $i = 1,2,\ldots$

To achieve more generality, here we give a different definition.

Definition 12.

a) *A finite series is a formal expression of the form* $\sum_{i=k}^{n} a_i$ *where* $\{a_i; i = k, k+1, \ldots, n\}$ *is a sequence of numbers.*

b) *An infinite series is a formal expression of the form* $\sum_{i=k}^{\infty} a_i$ *where* $\{a_i; i = k, k+1, k+2, \ldots\}$ *is an infinite sequence of numbers.*

c) *A standard finite series is a formal expression of the form* $\sum_{i=1}^{n} a_i$ *where* $\{a_i; i = 1, 2, \ldots, n\}$ *is a sequence of numbers.*

d) *A standard infinite series is a formal expression of the form* $\sum_{i=1}^{\infty} a_i$ *where* $\{a_i; i = 1, 2, \ldots\}$ *is an infinite sequence of numbers.*

This construction encompasses the definition given by Diedonne in [11] because both sequences $\{a_i; i = 1, 2, \ldots\}$ and $\{s_i; i = 1, 2, \ldots\}$ are uniquely and easily reconstructible from the expression $\sum_{i=1}^{\infty} a_i$. At the same time, the second sequence of partial sums from Diedonne's definition is superfluous in the case of finite series.

In real series, elements a_i are real numbers. In complex series, elements a_i are complex numbers. In hypernumber series, elements a_i are hypernumbers.

A series that has the form $\sum_{i=1}^{\infty} a$ has all elements equal to a. It is possible to consider finite series $\sum_{i=k}^{n} a_i$ and $\sum_{i=1}^{n} a_i$ as infinite series in which all elements with indices larger than n are equal to zero. Any natural summation of a finite series coincides with the standard addition of numbers in this series.

Our main concern here is infinite series (we call them simply series) as there are still many problems (both analytical and computational) with their summation, while finite series are known sufficiently well. For simplicity, we mostly study here operations with standard series as any series can be converted to a standard form. However, in some cases, results of operations depend on the form of series representation and we treat arbitrary series.

We consider three main classes of operations with series: *inner*, *intermediate*, and *outer*. *Inner* operations are performed with series as they are defined in this paper because the traditional approach does not make a clear distinction between a series and its sum. *Outer* operations are performed with sums of series. *Intermediate* operations are performed with partial sums of series. Many classical results demonstrate correspondences between inner, intermediate, and outer operations with series. For instance, it is known that the sum of the sum (inner operation) of two convergent series is equal to the sum (outer operation) of sums of these series. Traditionally, the sum of a series is defined as a limit (intermediate operation) of the partial sums of this series.

Inner operations are subdivided into two classes: *finite*, such as addition or multiplication, and *infinite*, such as summation.

3.1. Finite Operations on Series

Arithmetical operations with real numbers induce corresponding arithmetical operations in series.

Let $\sum_{i=1}^{\infty} a_i$ and $\sum_{i=1}^{\infty} b_i$ be series and c is a real number.

Definition 13. *Basic arithmetical operations on series:*

a) *Addition:*

$$\sum_{i=1}^{\infty} a_i + \sum_{i=1}^{\infty} b_i = \sum_{i=1}^{\infty} (a_i + b_i);$$

b) *Subtraction:*

$$\sum_{i=1}^{\infty} a_i - \sum_{i=1}^{\infty} b_i = \sum_{i=1}^{\infty} (a_i - b_i);$$

c) *Scalar multiplication:*

$$\left(\sum_{i=1}^{\infty} a_i\right) \cdot \left(\sum_{i=1}^{\infty} b_i\right) = \sum_{i=1}^{\infty} (a_i \cdot b_i);$$

d) *Scalar division:*

$$\left(\sum_{i=1}^{\infty} a_i\right) \Big/ \left(\sum_{i=1}^{\infty} b_i\right) = \sum_{i=1}^{\infty} (a_i / b_i)$$

assuming $b_i \neq 0$ *for all* $i = 1, 2, \ldots$;

e) *Multiplication by a number:*

$$c \cdot \left(\sum_{i=1}^{\infty} b_i\right) = \sum_{i=1}^{\infty} (c \cdot b_i);$$

f) *Additive multiplication:*

$$\left(\sum_{i=1}^{\infty} a_i\right) \cdot \left(\sum_{i=1}^{\infty} b_i\right) = \sum_{i=1}^{\infty} c_i,$$

where

$$c_i = \sum_{j+k=i} (a_j \cdot b_k).$$

By the definition, all laws of operations with real numbers (commutativity of addition, associativity of addition, commutativity of multiplication, associativity of multiplication, and distributivity) are valid for corresponding operations with series. Thus, we have the following result.

Proposition 3.1. *The space* $\mathbf{Ser}_{\mathbb{R}}$ $(\mathbf{Ser}_{\mathbb{C}})$ *of real (complex) series is a linear algebra over* $\mathbb{R}$ *with respect to addition, scalar multiplication, and multiplication by a real (complex) number.*

Basic operations allow one to define derived operations. Thus, any polynomial defines an arithmetical operation on series. For instance, the polynomial $p(x) = x^2 + 3x - 10$ defines the operation

$$\begin{aligned} p(x) : \sum_{i=1}^{\infty} a_i &\to \sum_{i=1}^{\infty} a_i^2 + 3 \cdot \left(\sum_{i=1}^{\infty} a_i\right) - 10 \\ &= \sum_{i=1}^{\infty} (a_i^2 + 3a_i) - 10 \\ &= a_1^2 + 3a_1 - 10 + \sum_{i=2}^{\infty} a_i^2 + 3 \cdot \left(\sum_{i=2}^{\infty} a_i\right). \end{aligned}$$

Definition 14. *A series* $\sum_{i=1}^{\infty} a_i$ *is a subseries of a series* $\sum_{i=1}^{\infty} b_i$ *if there is an injection* $\mathbf{j} : \mathbb{N} \to \mathbb{N}$ *such that* $a_i = b_{j(i)}$ *for all* $i = 1, 2, \ldots$.

A given injection $\mathbf{j}$ of natural numbers corresponds to a series $\sum_{i=1}^{\infty} b_i$ its subseries $\sum_{i=1}^{\infty} a_i$. This determines a unary *injection operation* on the set $\mathbf{Ser}_{\mathbb{R}}$ of all series. We denote this operation by $\sum_{i=1}^{\infty} a_i = \mathbf{j}\,(\sum_{i=1}^{\infty} b_i)$. This operation commutes with arithmetical operations.

Proposition 3.2. *For any series* $\sum_{i=1}^{\infty} a_i$ *and* $\sum_{i=1}^{\infty} b_i$ *and any injection* $\mathbf{j} : \mathbb{N} \to \mathbb{N}$, *we have:*

$$\mathbf{j}\left(\sum_{i=1}^{\infty} a_i\right) + \mathbf{j}\left(\sum_{i=1}^{\infty} b_i\right) = \mathbf{j}\left(\sum_{i=1}^{\infty} a_i + \sum_{i=1}^{\infty} b_i\right);$$

$$\mathbf{j}\left(\sum_{i=1}^{\infty} a_i\right) - \mathbf{j}\left(\sum_{i=1}^{\infty} b_i\right) = \mathbf{j}\left(\sum_{i=1}^{\infty} a_i - \sum_{i=1}^{\infty} b_i\right);$$

$$\mathbf{j}\left(\sum_{i=1}^{\infty} a_i\right) \cdot \mathbf{j}\left(\sum_{i=1}^{\infty} b_i\right) = \mathbf{j}\left(\sum_{i=1}^{\infty} a_i \cdot \sum_{i=1}^{\infty} b_i\right), \qquad \textit{and}$$

$$c \cdot \mathbf{j}\left(\sum_{i=1}^{\infty} b_i\right) = \mathbf{j}\left(c \cdot \sum_{i=1}^{\infty} b_i\right),$$

i.e., $\mathbf{j}$ *is an endomorphism of the linear algebra of all series.*

While the definition of a subseries is similar to the definition of subsequence, to introduce a quotient series, we need extra constructions.

Definition 15. *A projection* $\mathbf{p} : \mathbb{N} \to \mathbb{N}$ *is called:*

a) *cofinite if the set* $\mathbf{p}^{-1}(i)$ *is finite for all* $i = 1, 2, \ldots$;

b) *interval if* $\mathbf{p}^{-1}(i)$ *is a finite interval, i.e., has the form* $[k, h]$, *for all* $i = 1, 2, \ldots$

c) *monotone if* $i < j$ *implies* $\mathbf{p}(i) < \mathbf{p}(j)$ *for all* $i,\ j = 1, 2, \ldots$

Definition 16. *A series* $\sum_{i=1}^{\infty} a_i$ *is a (monotone) quotient series of a series* $\sum_{i=1}^{\infty} b_i$ *if there is an (monotone) interval projection* $\mathbf{p} : \mathbb{N} \to \mathbb{N}$ *such that* $a_i = \sum_{j \in p^{-1}(i)} b_j$ *for all* $i = 1, 2, \ldots$.

Informally, a quotient series is obtained from the initial series by addition of some elements from the initial series.

Remark 3.1. *It is possible to consider quotient series obtained from projections that glue together infinite subsets from* $\mathbb{N}$.

Given projection $\mathbf{p} : \mathbb{N} \to \mathbb{N}$ and a series $\sum_{i=1}^{\infty} b_i$, $\mathbf{p}$ corresponds a quotient series $\sum_{i=1}^{\infty} a_i$ to the series $\sum_{i=1}^{\infty} b_i$. This determines a unary *projection operation* on the set $\mathbf{Ser}_{\mathbb{R}}$ of all series. We denote this operation by $\sum_{i=1}^{\infty} a_i = \mathbf{p}\,(\sum_{i=1}^{\infty} b_i)$. This operation also commutes with arithmetical operations.

Proposition 3.3. *For any series $\sum_{i=1}^{\infty} a_i$ and $\sum_{i=1}^{\infty} b_i$ and any projection $\mathbf{p} : \mathbb{N} \to \mathbb{N}$, we have:*

$$\mathbf{p}\left(\sum_{i=1}^{\infty} a_i\right) + \mathbf{p}\left(\sum_{i=1}^{\infty} b_i\right) = \mathbf{p}\left(\sum_{i=1}^{\infty} a_i + \sum_{i=1}^{\infty} b_i\right);$$

$$\mathbf{p}\left(\sum_{i=1}^{\infty} a_i\right) - \mathbf{p}\left(\sum_{i=1}^{\infty} b_i\right) = \mathbf{p}\left(\sum_{i=1}^{\infty} a_i - \sum_{i=1}^{\infty} b_i\right);$$

$$c \cdot \mathbf{p}\left(\sum_{i=1}^{\infty} b_i\right) = \mathbf{p}\left(c \cdot \sum_{i=1}^{\infty} b_i\right),$$

i.e., $\mathbf{p}$ is an endomorphism of the linear space of all series.

However, $\mathbf{p}$ is not an endomorphism of the linear algebra of all series as $\mathbf{p}$ does not commute with multiplication.

The following example demonstrates how to build a quotient series.

Example 13. *Let $\sum_{i=1}^{\infty} b_i = 1 + \frac{1}{2} + \frac{1}{4} + \frac{1}{8} + \cdots$, i.e., $b_i = \frac{1}{2^{i-1}}$, and $\mathbf{p}(2n) = \mathbf{p}(2n-1) = n$. The projection $\mathbf{p}$ determines the monotone quotient series*

$$\sum_{i=1}^{\infty} a_i = \mathbf{p}\left(\sum_{i=1}^{\infty} b_i\right) = \frac{3}{2} + \frac{3}{8} + \frac{3}{32} + \cdots.$$

Definition 17. *A series $\sum_{i=1}^{\infty} a_i$ is a permutation of a series $\sum_{i=1}^{\infty} b_i$ if there is a permutation (bijection) $\mathbf{p} : \mathbb{N} \to \mathbb{N}$ such that $a_i = b_{p(i)}$ for all $i = 1, 2, \ldots$.*

Example 14. *Let $\sum_{i=1}^{\infty} b_i = 1 + 2 + 3 + 4 + 5 + \cdots$, i.e., $a_i = i$, and $\mathbf{p}(2n) = 2n - 1$, $\mathbf{p}(2n - 1) = 2n$. The permutation $\mathbf{p}$ determines the series*

$$\sum_{i=1}^{\infty} a_i = \mathbf{p}\left(\sum_{i=1}^{\infty} b_i\right) = 2 + 1 + 4 + 3 + 6 + 5 + \ldots.$$

3.2. Infinite Operations on Series

We start with the most popular and important infinite operation on series – summation. Opportunities for series opened by hypernumbers give us two kinds of summation: analytical and topological.

Definition 18. *The topological sum* $\mathrm{top}\sum_{i=1}^{\infty} a_i$ *of a standard series $\sum_{i=1}^{\infty} a_i$ is the limit of the sequence $\{S_n; n = 1, 2, \ldots\}$, where $S_n = a_1 + a_2 + \cdots + a_n$ for $n = 1, 2, 3, \ldots$ Traditionally, S_n is called a partial sum of the series $\sum_{i=1}^{\infty} a_i$.*

Definition 19. *A series $\sum_{i=1}^{\infty} a_i$ is called convergent (summable) in the classical sense if* $\mathrm{top}\sum_{i=1}^{\infty} a_i$ *exists.*

This is the classical definition of series summation. Mathematicians have found many properties of topological sums of numbers (cf., for example, [15, 9, 21]). For example, we have the following result.

Proposition 3.4. *For any series* $\sum_{i=1}^{\infty} a_i$*, if its topological sum* $\text{top}\sum_{i=1}^{\infty} a_i$ *exists, then it coincides with the topological sum of any of its monotone quotient series.*

Indeed, the classical limit of a sequence coincides with the classical limit of any its subsequences (cf., for example, [17]). If a series $\sum_{i=1}^{\infty} a_i$ is a monotone quotient series of a series $\sum_{i=1}^{\infty} b_i$, then the sequence $\{S_n; n = 1, 2, \ldots\}$ of the partial sums of the series $\sum_{i=1}^{\infty} a_i$ is a subsequence of the sequence $\{S_n^o; n = 1, 2, \ldots\}$ of the partial sums of the series $\sum_{i=1}^{\infty} b_i$. Consequently,

$$\text{top}\sum_{i=1}^{\infty} a_i = \lim S_n = \lim S_n^o = \text{top}\sum_{i=1}^{\infty} b_i.$$

Remark 3.2. *The Riemann series theorem* (cf. [13] or [1]) *shows that for non-monotone quotient series the result of Proposition 3.4 is not true.*

Definition 20. *A series* $\sum_{i=1}^{\infty} a_i$ *is called absolutely convergent (absolutely summable) in the classical sense if* $\text{top}\sum_{i=1}^{\infty} |a_i|$ *exists.*

Proposition 3.5. *Any absolutely convergent series is convergent.*

Results from [11, Ch. 5] imply the following property.

Proposition 3.6. *For any absolutely convergent series, its topological sum coincides with the topological sum of any of its quotient series.*

This property is, in a sense, a generalized associativity of absolutely convergent series because taking a quotient series means that some partial summation is done before the complete topological summation is performed.

At the same time, the Riemann series theorem (cf. [13] or [1]) shows that for conditionally convergent series the result of Proposition 3.6 is not true.

Definition 21 ([11]). *A series* $\sum_{i=1}^{\infty} a_i$ *is called commutatively convergent (summable) if* $\text{top}\sum_{i=1}^{\infty} a_i = \text{top}\sum_{i=1}^{\infty} b_i$ *for any permutation* $\sum_{i=1}^{\infty} b_i$ *of the series* $\sum_{i=1}^{\infty} a_i$.

Results from [11, Ch. 5] imply the following property.

Proposition 3.7. *A series is commutatively convergent if and only if it is absolutely convergent.*

In hypernumbers, we can give another definition for series summability.

Definition 22. *The analytical sum* $\text{an}\sum_{i=1}^{\infty} a_i$ *of a standard series* $\sum_{i=1}^{\infty} a_i$ *is equal to the hypernumber* $\text{Hn}(S_n)_{n\in\omega}$, *where* $S_n = a_1 + a_2 + \cdots + a_n$ *for* $n = 1, 2, 3, \ldots$

As any hypernumber is uniquely defined by any sequence, we have the following simple but important result.

Proposition 3.8. *For any series* $\sum_{i=1}^{\infty} a_i$*, its analytical sum always exists and is unique.*

Thus, any series has the sum. This sum is unique. The only difference in comparison with the standard approach to summability is that such a sum can be not only a real number but also an infinite hypernumber or/and oscillating hypernumber.

The notion of asymptotic expansions, which was introduced by Poincare in 1886, enables one to manipulate a large class of divergent series in much the same way as convergent power series [23]. Summation of series in hypernumbers allows one to manipulate all divergent series in much the same way as convergent series. As a consequence, it has many implications for asymptotic approximations of integrals.

It is important to know how the two considered types of summability are related.

Theorem 3.9. *For any series $\sum_{i=1}^{\infty} a_i$, its analytical sum coincides with its topological sum if and only if $an\sum_{i=1}^{\infty} a_i \in \mathbb{R}$.*

Proof. (1) Let $an\sum_{i=1}^{\infty} a_i \in \mathbb{R}$. Then

$$an\sum_{i=1}^{\infty} a_i = \mathrm{Hn}(S_n)_{n\in\omega} = a \in \mathbb{R}.$$

By the definition of a hypernumber, $\lim_{n\to\infty}|S_n - a| = 0$. Consequently, $\lim S_n = a$ and by the definition,

$$\mathrm{top}\sum_{i=1}^{\infty} a_i = a = an\sum_{i=1}^{\infty} a_i.$$

(2) Let

$$\mathrm{top}\sum_{i=1}^{\infty} a_i = an\sum_{i=1}^{\infty} a_i.$$

As $\mathrm{top}\sum_{i=1}^{\infty} a_i \in \mathbb{R}$, we have

$$an\sum_{i=1}^{\infty} a_i \in \mathbb{R}.$$

Theorem 3.9 is proved.

□

The result of Theorem 3.9 demonstrates that analytical summation is a natural extension of the classical summation. In other words, if a series has the sum in the classical sense, then the new extended definition gives the same sum for this series.

There are many properties and criteria that indicate when a series has the sum in the classical sense (cf., for example, [15, 9, 21]). Here we give new criteria for classical summability.

Theorem 3.9 implies the following result.

Corollary 3.10. *A series $\sum_{i=1}^{\infty} a_i$ is convergent in the classical sense if and only if $an\sum_{i=1}^{\infty} a_i \in \mathbb{R}$.*

Let us put $d_{n,m} = \sum_{i=n}^{m} a_i$.

Theorem 3.11. *A series $\sum_{i=1}^{\infty} a_i$ is convergent in the classical sense if and only if $\mathrm{Hn}(d_{j,n_j})_{j\in\omega} = 0$ for any sequence $\{n_j; j = 1, 2, \ldots\}$ with $n_j > j$.*

This property is equivalent to Cauchy criterion of series summability.

Proposition 3.12. *If a series $\sum_{i=1}^{\infty} a_i$ is convergent in the classical sense, then* $\mathrm{Hn}(a_i)_{i\in\omega} = 0$.

We have defined analytical summation only for standard series, in which enumeration starts with number 1 (cf. Definition 22). However, if we apply this definition to all series, the analytical summation would not inherit many useful properties of the classical (topological) summation. For instance, if $\sum_{i=1}^{\infty} a_i$ converges in the classical sense, then we have

$$\mathrm{top}\sum_{i=1}^{\infty} a_i = \mathrm{top}\sum_{i=1}^{n} a_i + \mathrm{top}\sum_{i=n}^{\infty} a_i$$

(additivity of summation) because for topological summation the number from which enumeration of elements starts does not change the topological sum. For analytical summation, this is not always true. Indeed, let us take $a_i = 1$ for all $i = 1, 2, \ldots$. Then $an\sum_{i=1}^{\infty} a_i = \mathrm{Hn}(i)_{i\in\omega}$, while

$$an\sum_{i=1}^{3} a_i + an\sum_{i=4}^{\infty} a_i = \mathrm{Hn}(3+i)_{i\in\omega} \neq \mathrm{Hn}(i)_{i\in\omega} = an\sum_{i=1}^{\infty} a_i.$$

In this example, the analytical sum of the series $\sum_{i=4}^{\infty} a_i$ is defined as $an\sum_{i=1}^{\infty} a_{i+3}$. Thus, to preserve additivity of summation, we define the analytical sum of series that start with i larger than 1 in a different way.

Definition 23. *The analytical sum $an\sum_{i=k}^{\infty} a_i$ of a series $\sum_{i=k}^{\infty} a_i$ is equal to the hypernumber* $\mathrm{Hn}(S_n)_{n\in\omega}$, *where $S_n = 0$ when $n < k$ and $S_n = a_k + a_{k+1} + \cdots + a_n$ when $n \geq k$.*

As a consequence of this definition and properties of hypernumbers, we have the same property as for the topological summation.

Proposition 3.13. *For any series $\sum_{i=1}^{\infty} a_i$ and any natural number n, we have*

$$an\sum_{i=1}^{\infty} a_i = an\sum_{i=1}^{n} a_i + an\sum_{i=n}^{\infty} a_i.$$

Summation in hypernumbers gives us new criteria for classical convergence considered in the next Theorem 3.15.

Lemma 3.14. *We have $an\sum_{i=n}^{\infty} a_i = an\sum_{i=1}^{\infty} b_i$ where $b_i = 0$ for $i < n$ and $b_i = a_i$ for $i \geq n$.*

Theorem 3.15. *The sum $an\sum_{i=k}^{\infty} a_i \in \mathbb{R}$ (or the series $\sum_{i=k}^{\infty} a_i$ converges in the classical sense) if and only if $an\sum_{i=k}^{\infty} a_i = an\sum_{i=k}^{\infty} b_i$ for any monotone quotient series $\sum_{i=k}^{\infty} b_i$ of the series $\sum_{i=k}^{\infty} a_i$.*

Proof. (1) Let $an\sum_{i=k}^{\infty} a_i \in \mathbb{R}$ and a series $\sum_{i=k}^{\infty} b_i$ be a monotone quotient series of the series $an\sum_{i=k}^{\infty} a_i$. By properties of hypernumbers, if the hypernumber defined by a sequence is a real number, then it coincides with the hypernumber defined by any subsequence of this sequence. If a series $\sum_{i=1}^{\infty} b_i$ is a monotone quotient series of a series $\sum_{i=1}^{\infty} a_i$, then the sequence $\{S_n; n = 1, 2, \ldots\}$ of the partial sums of the series $\sum_{i=1}^{\infty} b_i$ is a subsequence of the sequence $\{S_n^o; n = 1, 2, \ldots\}$ of the partial sums of the series $\sum_{i=1}^{\infty} a_i$. Consequently, $\mathrm{Hn}(S_n) = \mathrm{Hn}(S_n^o)$ or $an\sum_{i=k}^{\infty} a_i = an\sum_{i=k}^{\infty} b_i$.

(2) Let $an\sum_{i=k}^{\infty} a_i \notin \mathbb{R}$. Then there is a subsequence $(d_j)_{j\in\omega}$ of the sequence $(S_n)_{n\in\omega}$ of the partial sums S_n of the series $\sum_{i=k}^{\infty} a_i$ such that $an\sum_{i=k}^{\infty} a_i = \mathrm{Hn}(S_n)_{n\in\omega} \neq \mathrm{Hn}(d_j)_{j\in\omega}$. Indeed, in this case by Propositions 2.7 and 2.8, $\mathrm{Hn}\,(S_n)_{n\in\omega}$ is either infinite or oscillating. If $\mathrm{Hn}\,(S_n)_{n\in\omega}$ is an oscillating hypernumber, then its spectrum $\mathrm{Spec\,Hn}\,(S_n)_{n\in\omega}$ contains at least, two numbers. Let $\mathrm{Spec\,Hn}(S_n)_{n\in\omega} = \{a, b, \dots\}$. By the definition of spectrum, there is a subsequence $(d_j)_{j\in\omega}$ of the sequence $(S_n)_{n\in\omega}$ for which $\lim_{j\to\infty} d_j = a$. It means that $\mathrm{Spec\,Hn}(d_j)_{j\in\omega} = \{a\}$. Then $\mathrm{Hn}(S_n)_{n\in\omega} \neq \mathrm{Hn}(d_j)_{j\in\omega}$ as these hypernumbers have different spectra.

If the hypernumber $\mathrm{Hn}\,(S_n)_{n\in\omega}$ is infinite, then there is a subsequence $(b_i)_{i\in\omega}$ of the sequence $(S_n)_{n\in\omega}$ for which either $b_i \to \infty$ with $i \to \infty$ or $b_i \to -\infty$ with $i \to \infty$. Let us consider the first case. By the definition of $b_i \to \infty$ with $i \to \infty$, we can build a function $g(i)$ such that $b_{g(i)} > 2b_i$. Then by the definition of a hypernumber, $\mathrm{Hn}\,(b_i)_{i\in\omega} \neq \mathrm{Hn}(b_{g(i)})_{i\in\omega}$. Consequently, it is impossible that at the same time we have both equalities

$$\mathrm{Hn}(S_n)_{n\in\omega} = \mathrm{Hn}(b_i)_{i\in\omega} \quad \text{and} \quad \mathrm{Hn}(S_n)_{n\in\omega} = \mathrm{Hn}(b_{g(i)})_{i\in\omega}.$$

The case when $b_i \to -\infty$ with $i \to \infty$ is treated in a similar way.

To prove the theorem, we show that for any subsequence $(d_j)_{j\in\omega}$ of the sequence $(S_n)_{n\in\omega}$, there is a monotone quotient series $\sum_{i=k}^{\infty} b_i$ of the series $\sum_{i=k}^{\infty} a_i$ such that $(d_i)_{i\in\omega}$ is the sequence of partial sums of the series $\sum_{i=k}^{\infty} b_i$. Then we can find monotonous quotient series A and B of the series $\sum_{i=1}^{\infty} a_i$ such that their partial sums will be $\{b_i; n = 1, 2, 3, \dots\}$ and $\{b_{g(i)}; n = 1, 2, 3, \dots\}$, correspondingly.

Let us consider an arbitrary subsequence $(d_j)_{j\in\omega}$ of the sequence $(S_n)_{n\in\omega}$. We build the necessary monotone quotient series $\sum_{i=k}^{\infty} b_i$ by induction on the number j of the element d_j, constructing a monotonous projection $\mathbf{p}: \mathbb{N} \to \mathbb{N}$ (cf. Definition 15) such that $\sum_{i=k}^{\infty} b_i = \mathbf{p}\,(\sum_{i=k}^{\infty} a_i)$.

(1) If $j = 1$ and $d_1 = b_k$, then we determine that $\mathbf{p}$ combines k first members $a_1, \dots, a_k$ of the series $S = \sum_{i=1}^{\infty} a_i$ into one element $d_1 = a_1 + \dots + a_k$ of $\mathbf{p}S$.

(2) If $d_{j-1} = b_r$ and $d_j = b_t$. By the induction assumption, all elements of S are grouped by $\mathbf{p}$ so that the partial sums of $\mathbf{p}S$. are equal to d_l for all $l = 1, 2, \dots, j-1$. Then $b_t = b_r + a_{r+1} + a_{r+2} + \dots + a_t$. To get the necessary correspondence, we determine that $\mathbf{p}$ combines $a_{r+1}, a_{r+2}, \dots, a_t$ into one element $d_j = a_{r+1} + a_{r+2} + \dots + a_t$ of $\mathbf{p}S$. After this, all elements of S are grouped by $\mathbf{p}$ so that the partial sums of $\mathbf{p}S$ are equal to d_l for all $l = 1, 2, \dots, j$. This completes the inductive step.

By induction, we have that

$$\mathrm{Hn}(\mathbf{p}S) = \mathrm{Hn}(d_j)_{j\in\omega} \quad \text{and} \quad \mathrm{Hn}(S) = \mathrm{Hn}(S_n)_{n\in\omega} \neq \mathrm{Hn}(d_j)_{j\in\omega} = \mathrm{Hn}(\mathbf{p}S).$$

Theorem 3.15 is proved. □

Corollary 3.16. *If there is for any monotone quotient series $\sum_{i=k}^{\infty} b_i$ of the series $\sum_{i=k}^{\infty} a_i$ such that $an\sum_{i=k}^{\infty} a_i \neq an\sum_{i=k}^{\infty} b_i$, then the series $\sum_{i=k}^{\infty} a_i$ diverges in the classical sense.*

Similar to the sum of a series, it is possible to define the *inner product* of a series. Such a product also has two forms: analytical and topological.

Definition 24. *The topological inner product* $\text{top}\prod_{i=1}^{\infty} a_i$ *of a series* $\sum_{i=1}^{\infty} a_i$ *is the limit of the sequence* $\{P_n; n = 1, 2, \ldots\}$, *where* $P_n = a_1 \cdot a_2 \cdot \cdots \cdot a_n$ *for* $n = 1, 2, 3, \ldots$ *The number* P_n *is called a partial product of the series* $\sum_{i=1}^{\infty} a_i$.

Definition 25. *The analytical inner product* $\text{an}\prod_{i=1}^{\infty} a_i$ *of a series* $\sum_{i=1}^{\infty} a_i$ *is equal to the hypernumber* $\text{Hn}(P_n)_{n\in\omega}$, *where* $P_n = a_1 \cdot a_2 \cdot \cdots \cdot a_n$ *for* $n = 1, 2, 3, \ldots$

As $\log P_n = \log a_1 + \log a_2 + \cdots + \log a_n$, analytical and topological inner products of series are reduced to analytical and topological sums of series. In other words, we have identities:

$$an\left(\log \prod_{i=1}^{\infty} a_i\right) = an \sum_{i=1}^{\infty} \log a_i$$

and

$$\text{top}\left(\log \prod_{i=1}^{\infty} a_i\right) = \text{top} \sum_{i=1}^{\infty} \log a_i.$$

This allows one to deduce properties of inner products from properties of sums of a series. For instance, this reduction allows us to prove following results.

Proposition 3.17. *For any series* $\sum_{i=1}^{\infty} a_i$, *its analytical inner product always exists and is unique.*

Theorem 3.18. *For any series* $\sum_{i=1}^{\infty} a_i$, *its analytical inner product coincides with its topological inner product if and only if* $\text{an}\prod_{i=1}^{\infty} a_i \in \mathbb{R}$.

Theorem 3.19. $\text{an}\prod_{i=k}^{\infty} a_i \in \mathbb{R}$ *(or the infinite product* $\prod_{i=k}^{\infty} a_i$ *converges in the classical sense) if and only if* $\text{an}\sum_{i=k}^{\infty} \log a_i = \text{an}\sum_{i=k}^{\infty} b_i$ *for any monotone quotient series* $\sum_{i=k}^{\infty} b_i$ *of the series* $\sum_{i=k}^{\infty} \log a_i$.

It is possible to introduce infinite sums and products of series. However, we do not study these operations in this paper.

3.3. Outer and Intermediate Operations on Series

Outer operations are performed with sums of series. It means that to perform an outer operation, it is necessary to have two conditions:

(1) Sums of series under consideration exist.

(2) It is possible to perform corresponding operations with numerical values of these sums.

In the case of topological summation, usually the first condition is important because all operations that are considered are defined for real numbers. For instance, in the case of analytical summation, the situation is opposite: usually the second condition is important because analytical sums exist for all real series (cf. Proposition 3.8).

However, if we have to perform division top $\sum_{i=1}^{\infty} a_i \div$ top $\sum_{i=1}^{\infty} b_i$, it is necessary to verify that top $\sum_{i=1}^{\infty} b_i$ is not equal to zero. The first condition becomes essential for analytical summation in the case of infinite operation.

In many cases, outer operations are induced by inner operations on series. This is true for arithmetical operations on series that determine a vector/linear space $\mathbf{Ser}_{\mathbb{R}}$. Analytical sums of series in $\mathbb{R}_\omega$ preserve almost all properties of the classical (topological) sums of series in $\mathbb{R}$. For instance, we have the following result.

Proposition 3.20. *For any series $\sum_{i=1}^{\infty} a_i$ and $\sum_{i=1}^{\infty} b_i$, we have:*

$$an\sum_{i=1}^{\infty} a_i + an\sum_{i=1}^{\infty} b_i = an\left(\sum_{i=1}^{\infty} a_i + \sum_{i=1}^{\infty} b_i\right);$$

$$c\cdot\left(an\sum_{i=1}^{\infty} b_i\right) = an\left(c\cdot\sum_{i=1}^{\infty} b_i\right) \quad \textit{for any real number } c.$$

$$an\sum_{i=1}^{\infty} a_i - an\sum_{i=1}^{\infty} b_i = an\left(\sum_{i=1}^{\infty} a_i - \sum_{i=1}^{\infty} b_i\right).$$

Corollary 3.21. *Analytical summation is a homomorphism of the linear space* $\mathbf{Ser}_{\mathbb{R}}$ *of all real series into the linear space of all real hypernumbers.*

Corollary 3.22. *If series $\sum_{i=1}^{\infty} a_i$ and $\sum_{i=1}^{\infty} b_i$ are convergent in the classical sense, then:*

(1) the series $\sum_{i=1}^{\infty}(a_i + b_i)$ is convergent in the classical sense and

$$\text{top}\sum_{i=1}^{\infty} a_i + \text{top}\sum_{i=1}^{\infty} b_i = \text{top}\left(\sum_{i=1}^{\infty} a_i + \sum_{i=1}^{\infty} b_i\right);$$

(2) the series $c\cdot\sum_{i=1}^{\infty} b_i$ is convergent in the classical sense and

$$c\cdot\left(\text{top}\sum_{i=1}^{\infty} b_i\right) = \text{top}\left(c\cdot\sum_{i=1}^{\infty} b_i\right)$$

for any real number c.

(3) the series $\sum_{i=1}^{\infty}(a_i - b_i)$ is convergent in the classical sense and

$$\text{top}\sum_{i=1}^{\infty} a_i - \text{top}\sum_{i=1}^{\infty} b_i = \text{top}\left(\sum_{i=1}^{\infty} a_i - \sum_{i=1}^{\infty} b_i\right).$$

Corollary 3.23. *Topological summation is a homomorphism of the linear space of all convergent real series into the linear space of all real numbers.*

Remark 3.3. *For scalar and additive multiplications of series, these results are not always true because multiplication of hypernumbers is not always defined [4], while scalar and additive multiplications of series are always defined. To study analytical summation for scalar and additive multiplications of series, we need extended hypernumbers [6]. Here we do not consider series in the realm of extended hypernumbers.*

Nevertheless, we have somewhat weaker results in the context of hypernumbers.

Proposition 3.24. *If at least one of the hypernumbers* $an\sum_{i=1}^{\infty} a_i$ *or* $an\sum_{i=1}^{\infty} b_i$ *is bounded, then the product* $(an\sum_{i=1}^{\infty} a_i) \cdot (an\sum_{i=1}^{\infty} b_i)$ *exists and*

$$\left(an\sum_{i=1}^{\infty} a_i\right) \cdot \left(an\sum_{i=1}^{\infty} b_i\right) = an\left(\sum_{i=1}^{\infty} a_i \cdot \sum_{i=1}^{\infty} b_i\right).$$

Corollary 3.25. *If series* $\sum_{i=1}^{\infty} a_i$ *and* $\sum_{i=1}^{\infty} b_i$ *are convergent in the classical sense, then the product* $(\mathrm{top}\sum_{i=1}^{\infty} a_i) \cdot (\mathrm{top}\sum_{i=1}^{\infty} b_i)$ *exists and*

$$\left(\mathrm{top}\sum_{i=1}^{\infty} a_i\right) \cdot \left(\mathrm{top}\sum_{i=1}^{\infty} b_i\right) = \mathrm{top}\left(\sum_{i=1}^{\infty} a_i \cdot \sum_{i=1}^{\infty} b_i\right).$$

Intermediate operations are performed with partial sums of series before taking the sum of a series. For instance, we have the following definition for the analytical sum of the partial power of a series.

Definition 26. *The analytical sum of the partial power m of a standard series* $\sum_{i=1}^{\infty} a_i$ *is defined as*

$$an\left(\sum_{i=1}^{\infty} a_i\right)_P^m = \mathrm{Hn}(S_n^m)_{n\in\omega}$$

Definition 27. *The analytical sum of the partial outer product of standard series* $\sum_{i=1}^{\infty} a_i$ *and* $\sum_{i=1}^{\infty} b_i$ *is defined as*

$$an\left(\left(\sum_{i=1}^{\infty} a_i\right) \cdot_P \left(\sum_{i=1}^{\infty} b_i\right)\right) = \mathrm{Hn}(S_n \cdot R_n)_{n\in\omega},$$

where $S_n = a_1 + \cdots + a_n$ and $R_n = b_1 + \cdots + b_n$ for all $n = 1, 2, 3, \ldots$
As before, we have the following result.

Proposition 3.26.

a) *For any series* $\sum_{i=1}^{\infty} a_i$*, the analytical sum of its partial power m always exists and is unique.*

b) *For any series* $\sum_{i=1}^{\infty} a_i$ *and* $\sum_{i=1}^{\infty} b_i$*, its analytical sum of the partial outer product always exists and is unique.*

We have similar definitions for topological summation.

Definition 28. *The topological sum of the partial power m of a standard series $\sum_{i=1}^{\infty} a_i$ is defined as*

$$\operatorname{top}\left(\sum_{i=1}^{\infty} a_i\right)_P^m = \lim_{i\to\infty} S_n^m.$$

Definition 29. *The topological sum of the partial outer product of standard series $\sum_{i=1}^{\infty} a_i$ and $\sum_{i=1}^{\infty} b_i$ is defined as*

$$\operatorname{top}\left(\left(\sum_{i=1}^{\infty} a_i\right) \cdot_P \left(\sum_{i=1}^{\infty} b_i\right)\right) = \lim_{i\to\infty} (S_n \cdot R_n)$$

where $S_n = a_1 + \cdots + a_n$ and $R_n = b_1 + \cdots + b_n$ for all $n = 1, 2, 3, \ldots$

Remark 3.4. *It is possible that* $\operatorname{top}((\sum_{i=1}^{\infty} a_i) \cdot (\sum_{i=1}^{\infty} b_i))$ *exists, while neither* $\operatorname{top}\sum_{i=1}^{\infty} a_i$ *nor* $\operatorname{top}\sum_{i=1}^{\infty} b_i$ *are defined. For instance,* $\operatorname{top}\sum_{i=1}^{\infty} \frac{1}{i}$ *is not defined but*

$$\operatorname{top}\left(\left(\sum_{i=1}^{\infty} \frac{1}{i}\right) \cdot \left(\sum_{i=1}^{\infty} \frac{1}{i}\right)\right) = \operatorname{top}\sum_{i=1}^{\infty} \left(\frac{1}{i}\right)^2 = \frac{(\pi)^2}{6}.$$

4. Local and Global Inequalities for Series

Local inequalities show order relation between separate elements of series. We study how local inequalities influence properties of summation.

Local inequalities induce inequalities of series. Namely, we have the following definition.

Definition 30.

a) $\sum_{i=1}^{\infty} a_i \leq \sum_{i=1}^{\infty} b_i$ *for any series* $\sum_{i=1}^{\infty} a_i$ *and* $\sum_{i=1}^{\infty} b_i$ *if* $a_i \leq b_i$ *for all* $i = 1, 2, 3, \ldots$.

b) $\sum_{i=1}^{\infty} a_i < \sum_{i=1}^{\infty} b_i$ *for any series* $\sum_{i=1}^{\infty} a_i$ *and* $\sum_{i=1}^{\infty} b_i$ *if* $a_i < b_i$ *for all* $i = 1, 2, \ldots$.

Global inequalities reflect order relation between sums of series. Local inequalities often imply global inequalities.

Properties of real hypernumbers and finite sums of real numbers imply the following result.

Proposition 4.1. *If* $a_i \leq b_i$ *for all* $i = 1, 2, 3, \ldots$, *then* $\operatorname{an}\sum_{i=1}^{\infty} a_i \leq \operatorname{an}\sum_{i=1}^{\infty} b_i$ *for any series* $\sum_{i=1}^{\infty} a_i$ *and* $\sum_{i=1}^{\infty} b_i$.

Properties of limits of real numbers imply a similar result for the classical (topological) summation.

Proposition 4.2. *If* $\operatorname{top}\sum_{i=1}^{\infty} a_i$, $\operatorname{top}\sum_{i=1}^{\infty} b_i$ *exist for series* $\sum_{i=1}^{\infty} a_i$ *and* $\sum_{i=1}^{\infty} b_i$ *and* $a_i \leq b_i$ *for all* $i = 1, 2, \ldots$, *then* $\operatorname{top}\sum_{i=1}^{\infty} a_i \leq \operatorname{top}\sum_{i=1}^{\infty} b_i$.

Global inequalities influence operations with series that do not (absolutely) converge in $\mathbb{R}$. To prove our main result in this direction (Theorem 4.5), we need the two following lemmas.

Lemma 4.3. *A series $S = \sum_{i=1}^{\infty} a_i$ does not (absolutely) converge in $\mathbb{R}$ if and only if any its tail $\sum_{i=k}^{\infty} a_i$ does not (absolutely) converge in $\mathbb{R}$.*

Proof. Indeed, if $T = \sum_{i=k}^{\infty} a_i$ then by Proposition 4.1,

$$an \sum_{i=1}^{\infty} a_i = a_1 + \cdots + a_k + an \sum_{i=k}^{\infty} a_i = a + an \sum_{i=k}^{\infty} a_i$$

where a is some real number. In a similar way,

$$an \sum_{i=1}^{\infty} |a_i| = |a_1| + \cdots + |a_k| + an \sum_{i=k}^{\infty} |a_i| = b + an \sum_{i=k}^{\infty} |a_i|$$

where b is some positive real number. So, by Theorem 3.9, the series S (absolutely) converges in $\mathbb{R}$ if and only if the series T (absolutely) converges in $\mathbb{R}$. □

Lemma 4.4. *Any subsequence l of partial sums of the sequence of partial sums $\{S_n; n = 1, 2, \ldots\}$ of a series $\sum_{i=1}^{\infty} a_i$ determines a monotone quotient series $\sum_{i=1}^{\infty} d_i$ of the series $\sum_{i=1}^{\infty} a_i$ such that the sequence of partial sums of the series $\sum_{i=1}^{\infty} d_i$ coincides with l.*

Theorem 4.5 (The Bounded Riemann Series Theorem). *For any series $\sum_{i=1}^{\infty} a_i$, if there is a real number c such that $-c < an \sum_{i=1}^{\infty} a_i < c$ but $\sum_{i=1}^{\infty} a_i$ does not absolutely converge in the classical sense, then for any hypernumber α there exists a monotone quotient series $\sum_{i=1}^{\infty} d_i$ of the series $\sum_{i=1}^{\infty} a_i$, a permutation $\sum_{i=1}^{\infty} b_i$ of the series $\sum_{i=1}^{\infty} d_i$ and a monotone quotient series $\sum_{i=1}^{\infty} c_i$ of the series $\sum_{i=1}^{\infty} b_i$ such that $an \sum_{i=1}^{\infty} c_i = \alpha$.*

Proof. Let a series $\sum_{i=1}^{\infty} a_i$ be bounded, i.e., there is a real number c such that

$$-c < an \sum_{i=1}^{\infty} a_i < c,$$

but $\sum_{i=1}^{\infty} a_i$ does not absolutely converge in the classical sense, and $\alpha = \mathrm{Hn}(u_i)_{i\in\omega}$. Then the sequence $\{S_n; n = 1, 2, \ldots\}$ is bounded and thus, has a converging subsequence l. By Lemma 4.4, any subsequence of partial sums of the sequence of partial sums $\{S_n; n = 1, 2, \ldots\}$ determines a monotone quotient series of the series $\sum_{i=1}^{\infty} a_i$. The monotone quotient series $\sum_{i=1}^{\infty} d_i$ that corresponds to the converging subsequence l converges in $\mathbb{R}$ but does not absolutely converge in $\mathbb{R}$. Then by the Riemann series theorem, there is a permutation $\mathbf{q}_1$ of $\mathbb{N}$ and the corresponding permutation $\sum_{i=1}^{\infty} b_{1i}$ of the series $\sum_{i=1}^{\infty} a_i$ such that $\sum_{i=1}^{\infty} b_{1i}$ converges to the number u_1. By the definition of the convergence of series, there is some number k_1 such that the absolute value

$$\left| \sum_{i=1}^{m} b_{1i} - u_1 \right| < \frac{1}{2} \quad \text{for any } m > k_1.$$

The proof of the Riemann series theorem (cf. [13]) or the properties of convergent series show that it is possible to take such permutation $\mathbf{q}_1$ that changes places only for a finite number of elements from $\sum_{i=1}^{\infty} d_i$. So, there is the least number h_1 such that $\mathbf{q}_1$ does not move elements from $\sum_{i=1}^{\infty} d_i$ with indices larger than h_1.

Let $m_1 = \max\{k_1, h_1\}$ and $S_2 = \sum_{i=m_1+1}^{\infty} d_i$. As the permutation $\mathbf{q}_1$ does not change elements of S with larger than m_1 indices, S_2 is a tail of S. In addition, we have $\left|\sum_{i=1}^{m_1} b_i - u_1\right| < \frac{1}{2}$. By Lemma 4.3, the series S_2 converges in $\mathbb{R}$ but does not absolutely converge in $\mathbb{R}$. Then we can apply the same process to S_2. Namely, by the Riemann series theorem, there is a permutation $\mathbf{q}'_2$ of $\mathbb{N}$ and the corresponding permutation $\sum_{i=1}^{\infty} b_{2i}$ of the series S_2 such that $\mathbf{q}'_2 S_2 = \sum_{i=1}^{\infty} b_{2i}$ converges to the number a_2. By the definition of the convergence of series, there is some number k_2 such that

$$\left|\sum_{i=1}^{m} b_{2i} - u_2\right| < \frac{1}{4} \quad \text{for any } m > k_2.$$

The proof of the Riemann theorem (cf. [13]) or the properties of convergent series show that it is possible to take such permutation $\mathbf{q}_1$ that changes places only for a finite number of elements from S_2. So, there is the least number h_2 such that $\mathbf{q}'_2$ does not move elements from S_2 with larger indices.

Let $m_2 = \max\{k_2, h_2\}$ and $S_3 = \sum_{i=m_2+1}^{\infty} a_i$. As the permutation $\mathbf{q}'_2$ does not change elements of S_2 with larger than m_2 indices, S_3 is a tail of S_2 and thus, a tail of S. As the permutations $\mathbf{q}_1$ and $\mathbf{q}_2$ act at non-intersecting parts of the series S, we can define the permutation $\mathbf{q}_2 = \mathbf{q}_1 \cdot \mathbf{q}'_2$, which acts on S and changes the places only of those elements that have numbers less than or equal to m_2. In addition, we have

$$\left|\sum_{i=1}^{m_1} b_i - u_1\right| < \frac{1}{2} \quad \text{and} \quad \left|\sum_{i=m_1+1}^{m_2} b_{2i} - u_2\right| < \frac{1}{4}.$$

By Lemma 4.3, the series S_3 converges in $\mathbb{R}$ but does not absolutely converge in $\mathbb{R}$.

By induction on the number j of the element u_{j+1} from the sequence $(u_i)_{i\in\omega}$, it is possible to demonstrate that there are such a number $m_{j+1}$and a permutation $\mathbf{q}_{j+1}$ of $\mathbb{N}$ and the corresponding permutation $\sum_{i=1}^{\infty} b_{ji}$ of the series $\sum_{i=1}^{\infty} d_i$ for which the following conditions are satisfied. The permutation Q_{j+1} changes the places only of those elements from $\sum_{i=1}^{\infty} a_i$ that have numbers less than or equal to m_{j+1}. In addition, we have

$$\left|\sum_{i=1}^{m_1} b_{1i} - u_1\right| < \frac{1}{2}, \quad \left|\sum_{i=m_1+1}^{m_2} b_{2i} - u_2\right| < \frac{1}{4}, \quad \ldots,$$
$$\left|\sum_{i=m_j+1}^{m_{(j+1)}} b_{ji} - u_{j+1}\right| < \frac{1}{2^j}.$$

In addition, by Lemma 4.3, the series $S_{j+1} = \sum_{i=m_j+1}^{\infty} a_i$ converges in $\mathbb{R}$ but does not absolutely converge in $\mathbb{R}$.

Then the axiom of induction asserts that there is a permutation Q of the series $\sum_{i=1}^{\infty} d_i$ with the following property. There is a sequence of natural numbers $\{m_{j+1}.; j = 0, 1, 2, \ldots, n, \ldots\}$ such that

$$\left|\sum_{i=m_j+1}^{m_{(j+1)}} b_i - u_{j+1}\right| < \frac{1}{2^{j+1}} \qquad \text{for all} \quad j = 0, 1, 2, \ldots, n, \ldots.$$

Now we can take a partition P of the series QS such that $PQS = \sum_{j=1}^{\infty} c_j$, where $c_{j+1} = \sum_{i=m_j+1}^{m(j+1)} b_i$ for all $j = 0, 1, 2, \ldots, n, \ldots$. By the construction of the series PQS, we have $\mathrm{Hn}(PQS) = \alpha$.

Theorem 4.5 is proved. □

Definition 31. *A series $\sum_{i=k}^{\infty} a_i$ is called conditionally convergent in the classical sense if it converges in the classical sense but does not absolutely converge in the classical sense.*

The Riemann series theorem (cf. [13] or [1]) describes interesting properties of series that converge but do not converge absolutely. Namely, it states that by a suitable rearrangement of terms, a conditionally convergent series may be made to converge to any desired value, or to diverge. In the new setting of hypernumbers, when any series of numbers is convergent, it is possible to extend Riemann's result to the scope of all series.

Corollary 4.6 (The Generalized Riemann Series Theorem). *For any conditionally convergent, in the classical sense, series $\sum_{i=1}^{\infty} a_i$ and any hypernumber α there exists a permutation $\sum_{i=1}^{\infty} b_i$ of the series $\sum_{i=1}^{\infty} a_i$ and a monotone quotient series $\sum_{i=1}^{\infty} c_i$ of the series $\sum_{i=1}^{\infty} b_i$ such that $an \sum_{i=1}^{\infty} c_i = \alpha$.*

Corollary 4.7. *If a series $S = \sum_{i=1}^{\infty} a_i$ conditionally converges, in the classical sense, then for any series $T = \sum_{i=1}^{\infty} d_i$ there exists a permutation $\sum_{i=1}^{\infty} b_i$ of the series $\sum_{i=1}^{\infty} a_i$ and a monotone quotient series $\sum_{i=1}^{\infty} c_i$ of the series $\sum_{i=1}^{\infty} b_i$ such that $an \sum_{i=1}^{\infty} d_i = an \sum_{i=1}^{\infty} c_i$.*

Theorems 3.9 and 4.5 imply the following result.

Corollary 4.8 (The Riemann Series Theorem). *For any conditionally convergent, in the classical sense, series $\sum_{i=1}^{\infty} a_i$ and any real number a there is a permutation $\sum_{i=1}^{\infty} b_i$ of the series $\sum_{i=1}^{\infty} a_i$ such that* $\mathrm{top} \sum_{i=1}^{\infty} b_i = a$ *and there is a permutation $\sum_{i=1}^{\infty} c_i$ of the series $\sum_{i=1}^{\infty} a_i$ such that* $\mathrm{top} \sum_{i=1}^{\infty} c_i$ *tends to infinity.*

Theorem 4.5 allows us to find a new criterion for absolute convergence of series.

Theorem 4.9. *A series $\sum_{i=1}^{\infty} a_i$ absolutely converges, in the classical sense, if and only if $an \sum_{i=1}^{\infty} a_i = an \sum_{i=1}^{\infty} c_i$ for any monotone quotient series $\sum_{i=1}^{\infty} c_i$ of any permutation $\sum_{i=1}^{\infty} b_i$ of the series $\sum_{i=1}^{\infty} a_i$.*

Proof. Necessity of this condition follows from Theorem 3.9 and properties of absolutely convergent series (cf., Propositions 3.5 and 3.6).

Sufficiency. Let us consider a series $\sum_{i=1}^{\infty} a_i$ such that $an \sum_{i=1}^{\infty} a_i = an \sum_{i=1}^{\infty} c_i$ for any monotone quotient series $\sum_{i=1}^{\infty} c_i$ of any permutation $\sum_{i=1}^{\infty} b_i$ of the series $\sum_{i=1}^{\infty} a_i$. As this implies that $an \sum_{i=1}^{\infty} a_i = an \sum_{i=1}^{\infty} d_i$ for any monotone quotient series $\sum_{i=1}^{\infty} d_i$ of the series $\sum_{i=1}^{\infty} a_i$, by Theorem 4.5, $an \sum_{i=k}^{\infty} a_i \in \mathbb{R}$. By Theorem 3.9, $an \sum_{i=k}^{\infty} a_i = \mathrm{top} \sum_{i=k}^{\infty} a_i$, i.e., $\sum_{i=k}^{\infty} a_i$ converges in the classical sense. Then by Proposition 3.7, the series $\sum_{i=1}^{\infty} a_i$ absolutely converges in the classical sense.

Theorem 4.9 is proved. □

Let us consider a series $\sum_{i=1}^{\infty} a_i$ and the following condition:

$$an\sum_{i=1}^{\infty} a_i = an\sum_{i=1}^{\infty} b_i \tag{C}$$

for any permutation $\sum_{i=1}^{\infty} b_i$ of the series $\sum_{i=1}^{\infty} a_i$.

Remark 4.1. *In contrast to topological summation (cf. Proposition 3.6), Condition (C) does not imply absolute convergence of* $\sum_{i=1}^{\infty} a_i$.

Indeed, let us consider the series $\sum_{i=1}^{\infty} a_i$ in which all a_i are equal to 1. Then $an\sum_{i=1}^{\infty} a_i = an\sum_{i=1}^{\infty} b_i$ for any permutation $\sum_{i=1}^{\infty} b_i$ of this series. However, by the definition of a hypernumber, $an\sum_{i=1}^{\infty} a_i \notin \mathbb{R}$.

Local inequalities in series influence results of summation, both analytical and topological, determining topological convergence and types of analytical sums. For instance, Propositions 3.13, 4.2 and properties of increasing sequences of real numbers imply the following results.

Proposition 4.10. *If* $0 < a_i \leq b_i$ *for almost all* $i = 1, 2, \ldots$ *and* $an\sum_{i=1}^{\infty} b_i \in \mathbb{R}$, *then* $an\sum_{i=1}^{\infty} a_i \in \mathbb{R}$ *for any series* $\sum_{i=1}^{\infty} a_i$ *and* $\sum_{i=1}^{\infty} b_i$.

Corollary 4.11 (Comparison Test). *(cf. [22]). If* $0 < a_i \leq b_i$ *for all* $i = 1, 2, \ldots$ *and* $\sum_{i=1}^{\infty} b_i$ *converges, then* $\sum_{i=1}^{\infty} a_i$ *converges.*

Corollary 4.12. *If* $0 < a_i \leq b_i$ *for almost all* $i = 1, 2, \ldots$ *and* $an\sum_{i=1}^{\infty} a_i \notin \mathbb{R}$, *then* $an\sum_{i=1}^{\infty} b_i \notin \mathbb{R}$ *for any series* $\sum_{i=1}^{\infty} a_i$ *and* $\sum_{i=1}^{\infty} b_i$.

Proposition 4.13. *If* $0 < a_i \leq b_i$ *for almost all* $i = 1, 2, \ldots$ *and* $an\sum_{i=1}^{\infty} b_i$ *is bounded, then* $an\sum_{i=1}^{\infty} a_i$ *is bounded.*

Corollary 4.14. *If* $0 < a_i \leq b_i$ *for almost all* $i = 1, 2, \ldots$ *and* $an\sum_{i=1}^{\infty} a_i$ *is unbounded, then* $an\sum_{i=1}^{\infty} b_i$ *is unbounded for any series* $\sum_{i=1}^{\infty} a_i$ *and* $\sum_{i=1}^{\infty} b_i$.

Remark 4.2. *Condition "*$0 < a_i$ *for almost all* $i = 1, 2, \ldots$*" is necessary for the correctness of Propositions 4.10 and 4.13; as the following example demonstrates.*

Example 15. *Let*

$$\sum_{i=1}^{\infty} a_i = -1 + (-1) + (-1) + (-1) + \cdots \quad \textit{and}$$

$$\sum_{i=1}^{\infty} b_i = 1 + \frac{1}{2} + \frac{1}{4} + \frac{1}{8} + \cdots,$$

i.e., $a_i = \frac{1}{2^{i-1}}$. *Then* $an\sum_{i=1}^{\infty} b_i = 2$, *while the hypernumber* $an\sum_{i=1}^{\infty} a_i$ *is infinite and* $a_i \leq b_i$ *for all* $i = 1, 2, \ldots$.

Proposition 4.15. *If* $0 \le a_i$, $0 \le b_i$ *for almost all* $i = 1,2,\ldots$, $\mathrm{Hn}(a_i/b_i)_{i\in\omega} \ne 0$ *and an* $\sum_{i=1}^{\infty} b_i \in \mathbb{R}$, *then an* $\sum_{i=1}^{\infty} a_i \in \mathbb{R}$ *for any series* $\sum_{i=1}^{\infty} a_i$ *and* $\sum_{i=1}^{\infty} b_i$.

Proposition 4.16. *If* $0 \le a_i$, $0 \le b_i$ *for almost all* $i = 1,2,\ldots$, $\mathrm{Hn}(a_i/b_i)_{i\in\omega} \ne 0$ *and an* $\sum_{i=1}^{\infty} b_i$ *is bounded, then an* $\sum_{i=1}^{\infty} a_i$ *is bounded, for any series* $\sum_{i=1}^{\infty} a_i$ *and* $\sum_{i=1}^{\infty} b_i$.

Let $\mathrm{Hn}(a_{i+1}/a_i)_{i\in\omega} = \alpha$ and all a_i be nonnegative numbers.

Theorem 4.17.

a) If α *is a bounded hypernumber and there is a number* $k > 0$ *such that for any* $a \in \mathrm{Spec}\,\alpha$, *we have* $0 \le a < 1-k$, *then an* $\sum_{i=1}^{\infty} a_i \in \mathbb{R}$.

b) If α *is an infinite hypernumber, then an* $\sum_{i=1}^{\infty} a_i \notin \mathbb{R}$.

c) If there is a number $k > 0$ *such that for any* $a \in \mathrm{Spec}\,\alpha$, *we have* $a > 1+k$, *then an* $\sum_{i=1}^{\infty} a_i$ *is an increasing infinite hypernumber, in particular, an* $\sum_{i=1}^{\infty} a_i \notin \mathbb{R}$.

Proof. a) Let us assume that there is a number $k > 0$ such that for any $a \in \mathrm{Spec}\,\alpha$, we have $a < 1-k$. Denote a_{i+1}/a_i by b_i. Then for any number $h > 0$, there are only a finite number of elements b_i that are larger than or equal to $1-k+h$. Indeed, if there are infinitely many numbers b_i that are larger than or equal to $1-k+h$, then the sequence of all these numbers either has a subsequence that goes to infinity or, as the real line is a complete metric space, the sequence $\{b_i; i = 1,2,3,\ldots\}$ has a converging subsequence l. Both cases violate the initial condition because in the first case, it contradicts the boundedness of the hypernumber α, while in the second case, the limit c of the subsequence l belongs to $\mathrm{Spec}\,\alpha$ and is larger than or equal to $1-k+h$ and thus, larger than $1-k$.

Taking $h = \frac{1}{2}k$, we have that there is only a finite number of elements b_i that are larger than or equal to $1-\frac{1}{2}k$. Denote $1-\frac{1}{2}k$ by q. Then $a_{i+1}/a_i < q$ for almost all $i = 1,2,\ldots$. Consequently, $a_{i+1} < q \cdot a_i$ for almost all $i = 1,2,\ldots$.

Let us take the least number n such that $a_{i+1} < q \cdot a_i$ for all $i = n, n+1, n+2, \ldots$ and show that *an* $\sum_{i=1}^{\infty} d_i \in \mathbb{R}$, where $d_i = a_{i+n-1}$ for all $i = 1,2,\ldots$. Indeed, denote d_1 by d. Then $d_1 = d$, $d_2 < q \cdot d$, $d_3 < q^2 \cdot d$, $d_4 < q^3 \cdot d$, and so on. By induction, we prove that $d_i < q^{i-1} \cdot d$. As $q < 1$, the series $\sum_{i=1}^{\infty} q^i \cdot d$ converges in the classical sense [22]. Thus, by Theorem 3.9, *an* $\sum_{i=1}^{\infty} q^i \cdot d \in \mathbb{R}$. Then by Proposition 4.10, *an* $\sum_{i=1}^{\infty} d_i \in \mathbb{R}$ and *an* $\sum_{i=1}^{\infty} a_i \in \mathbb{R}$.

Part a) of Theorem 4.17 is proved.

b) If α is an infinite hypernumber, then the sequence $\{b_i; i = 1,2,\ldots\}$ is unbounded. Consequently, the sequence $\{a_i; i = 1,2,\ldots\}$ is unbounded. Then by Proposition 3.12, *an* $\sum_{i=1}^{\infty} a_i \notin \mathbb{R}$.

c) Let us assume that there is a number $k > 0$ such that for any $a \in \mathrm{Spec}\,\alpha$, we have $a > 1+k$. Denote a_{i+1}/a_i by b_i. Then for any number $h > 0$, there are only a finite number of elements b_i that are smaller than or equal to $1+k-h$. Indeed, if there are infinitely many numbers b_i that are smaller than or equal to $1+k-h$, then the set of all these numbers has a converging subsequence l as the real line is a complete metric space. The limit c of this subsequence l belongs to $\mathrm{Spec}\,\alpha$ and is smaller than or equal to $1+k-h$ and thus, smaller than $1+k$. This contradicts the initial condition of part c).

Taking $h = \frac{1}{2}k$, we have that there are only infinitely many numbers b_i that are larger than or equal to $1 + \frac{1}{2}k$. Denote $1 + \frac{1}{2}k$ by q. Then $a_{i+1}/a_i > q$ for almost all $i = 1, 2, \ldots$. Consequently, $a_{i+1} > q \cdot a_i$ for almost all $i = 1, 2, \ldots$.

Let us take the least number n such that $a_{i+1} > q \cdot a_i$ for all $i = n, n+1, n+2, \ldots$ and show that $an\sum_{i=1}^{\infty} d_i \in \mathbb{R}$, where $d_i = a_{i+n-1}$ for all $i = 1, 2, \ldots$. Indeed, denote d_1 by d. Then $d_1 = d$, $d_2 > q \cdot d$, $d_3 > q^2 \cdot d$, $d_4 > q^3 \cdot d$, and so on. By induction, we prove that $d_i > q^{i-1} \cdot d$. As $q > 1$, the series $\sum_{i=1}^{\infty} d_i$ diverges in the classical sense [22]. Thus, by Theorem 3.9 and Proposition 3.12, $an\sum_{i=1}^{\infty} a_i \notin \mathbb{R}$.

Theorem 4.17 is proved. □

Corollary 4.18.

a) If $a_{i+1}/a_i < 1$ for almost all $i = 1, 2, \ldots$, then $\sum_{i=1}^{\infty} a_i$ converges in the classical sense.

b) If $a_{i+1}/a_i > 1$ for almost all $i = 1, 2, \ldots$, then $\sum_{i=1}^{\infty} a_i$ diverges in the classical sense.

As in the classical case (cf., for example, [22]), we have that if, at least, one element a from $\operatorname{Spec} \alpha$ is equal to 1, then the corresponding series $\sum_{i=1}^{\infty} a_i$ may be converging or be diverging.

Remark 4.3. *For divergence of the series $\sum_{i=1}^{\infty} a_i$ in Part c) of Theorem 4.17, it is not sufficient to have the inequality $a > 1 + k$ for all but one of the elements from* $\operatorname{Spec} \alpha$ *as the following example demonstrates.*

Example 16. *Let us consider the series $\sum_{i=1}^{\infty} a_i$ in which $a_{2i} = (\frac{1}{2})^{2i-1}$ and $a_{2i-1} = (\frac{1}{2})^{2i}$ for $i = 1, 2, 3, \ldots$. This series has the expanded form $\frac{1}{4} + \frac{1}{2} + \frac{1}{16} + \frac{1}{8} + \frac{1}{64} + \frac{1}{32} + \cdots$. Then $Hn(c_i)_{i\in\omega} = \alpha$ where $c_i = \frac{a_{i+1}}{a_i}$, i.e., $c_1 = 2$, $c_2 = \frac{1}{8}$, $c_3 = 2$, $c_4 = \frac{1}{8}$, and consequently,* $\operatorname{Spec} = \{2, \frac{1}{8}\}$. *At the same time,*

$$an\sum_{i=1}^{\infty} a_i = an\sum_{i=1}^{\infty} \left(\frac{1}{2}\right)^i = an\left(\frac{1}{2} + \left(\frac{1}{2}\right)^2 + \left(\frac{1}{2}\right)^3 + \cdots\right) = 1,$$

i.e., the series $\sum_{i=1}^{\infty} a_i$ converges in the classical sense.

It is possible to get a criterion for classical convergence in terms of global inequalities.

Theorem 4.19. *A series $\sum_{i=k}^{\infty} a_i$ converges in the classical sense if and only if*

$$an\sum_{i=k}^{\infty} a_i \geq an\sum_{i=k}^{\infty} b_i$$

for any monotone quotient series $\sum_{i=k}^{\infty} b_i$ of the series $\sum_{i=k}^{\infty} a_i$.

Proof. The proof is similar to the proof of Theorem 3.15 and will be omitted. □

5. Inequalities and Operations on Series

There are many important inequalities for topological sums of series. The Cauchy inequality, Hölder's inequality and its generalizations, Minkowsky's inequality, and Chebyshev's inequality [14] are the most popular examples. Analytical summation essentially extends topological summation. Here we prove some of these inequalities for analytical sums of series.

For finite series (sums), Cauchy's inequality (also called the Cauchy-Schwarz inequality or Lagrange's inequality) has the following form:

$$\left(\sum_{i=1}^{n} a_i b_i\right)^2 \leq \left(\sum_{i=1}^{n} a_i^2\right)\left(\sum_{i=1}^{n} b_i^2\right)$$

where equality holds if and only if $a_i = b_i$ for all $i = 1,2,3,\ldots$

It is a special case of Hölder's inequality for finite sums with $p = q = 2$ (cf., [14]). This inequality arises in many fields, so it is known under several other names, for example, as the Buniakovsky inequality or the Kantorovich inequality.

Theorem 5.1 (Cauchy inequality for analytical sums). *For arbitrary series* $\sum_{i=1}^{\infty} a_i$ *and* $\sum_{i=1}^{\infty} b_i$, *we have*

$$an\left(\sum_{i=1}^{\infty} a_i b_i\right)_P^2 \leq an\left(\left(\sum_{i=1}^{\infty} a_i^2\right) \cdot_P \left(\sum_{i=1}^{\infty} b_i^2\right)\right). \tag{5.1}$$

The inequality (5.1) becomes an equality if and only $\sum_{i=1}^{\infty} a_i = c \cdot (\sum_{i=1}^{\infty} b_i)$ for some real number c.

Proof. One way to prove the Cauchy inequality for finite series is to consider the finite series

$$\frac{1}{2}\sum_{i=1}^{n}(a_i b_j - a_j b_i)^2. \tag{5.2}$$

On the one hand, the series (5.2) is nonnegative and when, at least, $(a_i/b_i) \neq (a_j/b_j)$, the series (5.2) is positive. On the other hand, we have

$$\frac{1}{2}\sum_{i=1}^{n}(a_i b_j - a_j b_i)^2 = \left(\sum_{i=1}^{n} a_i^2\right) \times \left(\sum_{i=1}^{n} b_i^2\right) - \sum_{i=1}^{n}(a_i b_i)^2,$$

where $\times$ is the multiplication of real numbers.

Thus,

$$\left(\sum_{i=1}^{n} a_i^2\right) \times \left(\sum_{i=1}^{n} b_i^2\right) - \sum_{i=1}^{n}(a_i b_i)^2 \geq 0$$

or

$$\sum_{i=1}^{n}(a_i b_i)^2 \leq \left(\sum_{i=1}^{n} a_i^2\right) \times \left(\sum_{i=1}^{n} b_i^2\right).$$

Thus,

$$an\left(\sum_{i=1}^{\infty} a_i b_i\right) p^2 \leq an\left(\left(\sum_{i=1}^{\infty} a_i^2\right) \cdot_P \left(\sum_{i=1}^{\infty} b_i^2\right)\right)$$

as the number k cannot decrease with the growth of n.

By Definitions 26 and 27,

$$an\left(\sum_{i=1}^{\infty} a_i b_i\right)_P^2 = \mathrm{Hn}\left(\sum_{i=1}^{n} (a_i b_i)^2\right)_{n\in\omega}$$

and

$$an\left(\left(\sum_{i=1}^{\infty} a_i^2\right) \cdot_P \left(\sum_{i=1}^{\infty} b_i^2\right)\right) = \mathrm{Hn}\left(\left(\sum_{i=1}^{n} a_i^2\right) \times \left(\sum_{i=1}^{n} b_i^2\right)\right)_{n\in\omega}$$

By the definition of order for hypernumbers, we have:

$$\text{if } \alpha = \mathrm{Hn}(d_n))_{n\in\omega},\ \beta = \mathrm{Hn}(c_n))_{n\in\omega}, \quad \text{and}$$
$$d_n \leq c_n \quad \text{for all } i = 1, 2, 3, \ldots, \quad \text{then} \quad \alpha \leq \beta.$$

Consequently,

$$an\left(\sum_{i=1}^{\infty} a_i b_i\right)_P^2 \leq an\left(\left(\sum_{i=1}^{\infty} a_i^2\right) \cdot_P \left(\sum_{i=1}^{\infty} b_i^2\right)\right).$$

Besides, when $\sum_{i=1}^{\infty} a_i = c \cdot (\sum_{i=1}^{\infty} b_i)$ for some real number c, all differences $a_i b_j - a_j b_i$ and thus, the series (5.2) are equal to zero. Consequently, we see that the condition in the second part of the theorem is sufficient.

To prove necessity of this condition, we note that partial sums of the series $\frac{1}{2}\sum_{i=1}^{\infty}(a_i b_j - a_j b_i)^2$ are strictly increasing. Thus, if at least one difference $a_i b_j - a_j b_i$ is not equal to zero, then for some n and all larger numbers, we have

$$\frac{1}{2}\sum_{i=1}^{n}(a_i b_j - a_j b_i)^2 \geq \frac{1}{2}(a_i b_j - a_j b_i)^2 = k > 0.$$

Consequently,

$$\left(\sum_{i=1}^{n} a_i^2\right) \times \left(\sum_{i=1}^{n} b_i^2\right) - \sum_{i=1}^{n}(a_i b_i)^2 \geq k$$

and

$$\left(\sum_{i=1}^{n} a_i^2\right) \times \left(\sum_{i=1}^{n} b_i^2\right) \geq \sum_{i=1}^{n}(a_i b_i)^2 + k.$$

Consequently,

$$\left(\sum_{i=1}^{n} a_i^2\right) \times \left(\sum_{i=1}^{n} b_i^2\right) > \sum_{i=1}^{n}(a_i b_i)^2.$$

The theorem is proved. □

Theorems 3.9 and 5.1 imply the following result.

Corollary 5.2 (Cauchy inequality for topological sums). *If the series* $\sum_{i=1}^{\infty} a_i^2$ *and* $\sum_{i=1}^{\infty} b_i^2$ *are convergent in the classical sense, then*

$$\text{top}\left(\sum_{i=1}^{\infty} a_i b_i\right)_P^2 \leq \text{top}\left(\sum_{i=1}^{\infty} a_i^2\right) \cdot \text{top}\left(\sum_{i=1}^{\infty} b_i^2\right). \tag{5.3}$$

The inequality (5.3) becomes an equality if and only $\sum_{i=1}^{\infty} a_i = c \cdot (\sum_{i=1}^{\infty} b_i)$ *for some real number c.*

Using the Hölder sum inequality, we can prove the following result.

Theorem 5.3 (Hölder inequality for analytical sums). *For arbitrary series* $\sum_{i=1}^{\infty} a_i$ *and* $\sum_{i=1}^{\infty} b_i$*, we have*

$$an \sum_{i=1}^{\infty} |a_i b_i| \leq an\left(\left(\sum_{i=1}^{\infty} |a_i|^p\right)_P^{\frac{1}{p}}\right) \cdot_P \left(\left(\sum_{i=1}^{\infty} |b_i|^q\right)_P^{\frac{1}{q}}\right), \tag{5.4}$$

where $\frac{1}{p} + \frac{1}{q} = 1$*. The inequality (5.4) becomes an equality if and only* $\sum_{i=1}^{\infty} a_i = c \cdot (\sum_{i=1}^{\infty} b_i)$ *for some real number c.*

6. Conclusion

In mathematics, a large number of methods have been proposed for the summation of divergent series: Cesàro summation, Abel summation, Borel summation, etc. All these methods define the sum of a series as a real number but do not allow one to find a sum of many series. Extending numbers to hypernumbers, we can find the sum of any series. As is demonstrated in this paper, this summation has many useful properties similar to properties of conventional summation methods [18]. Generalized summation is usually built in the form of some linear functional L with domain contained in some space S of numerical sequences. This implies a possibility to extend existing, or at least, the most popular methods of summation to hypernumbers. It would be interesting to consider the following problem. To build such an extension directly as it is done for classical summation in the definition of the analytical sum of a series or to use a general technique for extending linear functionals developed in [2] for hypernumbers and extrafunctions and to study inequalities for generalized analytical summation.

One more interesting problem is to build fuzzy summation of series in the sense of fuzzy limits [3, 7] and to study inequalities for fuzzy topological summation.

Bibliography

[1] T.J. Bromwich, and T.M. MacRobert, *An Introduction to the Theory of Infinite Series*, Chelsea, New York, 1991.

[2] M. Burgin, On the Hahn-Banach theorem for hyperfunctionals, *Notices of the Academy of Sciences of Ukraine*, **7** (1991), 9–14 (in Russian and Ukrainian).

[3] M. Burgin, Theory of fuzzy limits, *Fuzzy Sets and Systems*, **115**(3) (2000), 433–443.

[4] M. Burgin, Theory of hypernumbers and extrafunctions: functional spaces and differentiation, *Discrete Dynamics in Nature and Society*, **7** (2002), 201–212.

[5] M. Burgin, Hyperfunctionals and Generalized Distributions, in "*Stochastic Processes and Functional Analysis*" (Eds. Krinik, A.C. and Swift, R.J.), A Dekker Series of Lecture Notes in Pure and Applied Mathematics, v. 238, (2004), 81–119.

[6] M. Burgin, Topology in nonlinear extensions of hypernumbers, *Discrete Dynamics in Nature and Society*, **10**(2) (2005), 145–170.

[7] M. Burgin and M. Kalina, Fuzzy conditional convergence and nearness relations, *Fuzzy Sets and Systems*, **149**(3) (2005), 383–398.

[8] M. Burgin and J. Ralston, PDE and extrafunctions, *Rocky Mountain Journal of Mathematics*, **34** (2004), 849–867.

[9] T. Davis, *The Summation of Series*, San Antonio, Principia Press of Trinity University, 1962.

[10] P. Deligne, P. Etingof, D.S. Freed, L.C. Jeffrey, D. Kazhdan, J.W. Morgan, D.R. Morrison, and E. Witten (Eds.), *Quantum Fields and Strings: A Course for Mathematicians*, American Mathematical Society, Providence, RI, 1999.

[11] J. Diedonne, *Foundations of Modern Analysis,* New York / London, Academic Press, 1960.

[12] H. Edzawa and T. Zuneto (Eds.), *Quantum Physics Perspectives,* Iwanami Shoten, Tokio, 1977.

[13] G.M. Fihtengoltz, *Elements of Mathematical Analysis*, Moscow, GITTL, 1955 (in Russian).

[14] G.H. Hardy, J.E. Littlewood and G. Pólya, *Inequalities*, 1934.

[15] L.B.W. Jolley, *Summation of Series*, Dover Pubns, 1961.

[16] A. Robinson, *Non-Standard Analysis*, Studies of Logic and Foundations of Mathematics, New York, North-Holland, 1966.

[17] K.A. Ross, *Elementary Analysis: The Theory of Calculus,* New York / Berlin / Heidelberg, Springer, 1996.

[18] C. Rousseau, *Divergent Series: Past, Present, Future* ..., http://www.dms.umontreal.ca/~rousseau/.

[19] L. Schwartz, *Theorie de distributions*, Vol. I-II, Hermann, Paris, 1950-1951.

[20] W. Siegel, *Introduction to String Field Theory*, World Scientific, Singapore, 1988.

[21] A. Sofo, *Computational Techniques for the Summation of Series*, Springer, 2003.

[22] J. Stuart, *Calculus,* Pacific Grove, California, Brooks/Cole P.C., 1991.

[23] R. Wang, *Asymptotic Approximations of Integrals,* Academic Press, Inc., Boston/San Diego/New York, 1989.

In: Advances in Inequalities for Series
Editors: S. S. Dragomir and A. Sofo, pp. 121–129
ISBN 978-1-60021-920-7

Chapter 6

A New Class of Salagean-Type $p-$valent Functions

H. Özlem Güney
University of Dicle, Faculty of Science & Arts
Department of Mathematics, 21280 Diyarbakır-TURKEY
E-mail address: ozlemg@dicle.edu.tr

Abstract

The object of the present paper is to derive several interesting properties of the subclass $T_p^*(g(z),\alpha,\beta,\gamma,A,B,\lambda)$ consisting of analytic and $p-$valent functions with negative coefficients. We prove coefficient inequalities and distortion theorems with the help of the Salagean differential operator D^n. Also, extreme points for $T_p^*(g(z),\alpha,\beta,\gamma,,A,B,\lambda)$ are determined.

1991 Mathematics Subject Classification: 30C45

Key words and phrases: Analytic, $p-$valent, Salagean operator, extreme points.

1. Introduction

Let A_p $(p \geq 2)$ denote the class of functions of the form

$$f(z) = z^p + \sum_{k=1}^{\infty} a_{p+k} z^{p+k} \tag{1.1}$$

which are analytic and $p-$valent in the unit disk.

Let T_p^* denote the subclass of A_p whose members have the form

$$f(z) = z^p - \sum_{k=1}^{\infty} a_{p+k} z^{p+k}, \quad a_{p+k} \geq 0. \tag{1.2}$$

We can write the following equalities for the functions $f(z)$ belonging to the class T_p^*:

$$D^0 f(z) = f(z)$$

$$D^1 f(z) = Df(z) = zf'(z) = pz^p - \sum_{k=1}^{\infty} (p+k)a_{p+k}z^{p+k},$$

$$\vdots$$

$$D^\lambda f(z) = D(D^{\lambda-1} f(z)) = p^\lambda z^p - \sum_{k=1}^{\infty} (p+k)^\lambda a_{p+k}z^{p+k}.$$

The differential operator D^λ was introduced by Salagean [5]. With the help of the differential operator D^λ, a function $f \in T_p^*(g(z), \alpha, \beta, \gamma, A, B, \lambda)$ if and only if

$$\left| \frac{z(D^\lambda f(z))' - pD^\lambda g(z)}{(A-B)(p-\beta)D^\lambda g(z) - B\left[z(D^\lambda f(z))' - pD^\lambda g(z)\right]} \right| < \gamma, \quad z \in U \tag{1.3}$$

for $g(z) \in T_p^*(\alpha)$, $-1 \le B < A \le 1$, $-1 \le B < 0$, $0 \le \alpha < p$, $0 \le \beta < p$ and $0 < \gamma < 1$.

We note that for $p = 1, A = 1$, $B = -1$ and $\lambda = 0$, the class $T_p^*(g(z), \alpha, \beta, \gamma, 1, -1, 0)$ reduces to the class $T_p^*(g(z), \alpha, \beta, \gamma)$ which was introduced and studied by Srivastava and Owa [2]. Further, for $\alpha = p = 1$, we get $g(z) \equiv z$ so that for $A = 1$, $B = -1$ and $\lambda = 0$, $T_p^*(z, 1, \beta, \gamma, 1, -1, 0)$ gives the class $P^*(\beta, \gamma)$ considered by Gupta and Jain [3] which in turn leads to the class studied by Sarangi and Uralegaddi [4] for $\alpha = 1$. A function $f \in T_p^*$ is in $T_p^*(g(z), \alpha, \beta, \gamma, A, B, 0)$ if and only if

$$\left| \frac{\frac{zf'(z)}{g(z)} - p}{(A-B)(p-\beta) - B\left(\frac{zf'(z)}{g(z)} - p\right)} \right| < \gamma,$$

for $g(z) \in T_p^*(\alpha)$, $-1 \le B < A \le 1$, $-1 \le B < 0$, $0 \le \alpha < p$, $0 \le \beta < p$ and $0 < \gamma < 1$. The class $T_p^*(g(z), \alpha, \beta, \gamma, A, B, 0)$ was investigated by Patel and Rout [1]. In the present paper, we generalize the results of Patel and Rout [1].

2. Coefficient Theorem

We need the following lemma.

Lemma 2.1. *Let the function $g(z)$ be defined by $g(z) = z^p - \sum_{k=1}^{\infty} b_{p+k}z^{p+k}$. Then $g(z)$ belongs to the class $T_p^*(\alpha, \lambda)$ if and only if*

$$\sum_{k=1}^{\infty} (p+k)^\lambda (p+k-\alpha) b_{p+k} \le p^\lambda (p-\alpha). \tag{2.1}$$

Theorem 2.2. *Let the function $f(z)$ defined by (1.2) be in the class $T_p^*(g(z), \alpha, \beta, \gamma, A, B, \lambda)$. Then*

$$\sum_{k=1}^{\infty} (p+k)^{\lambda+1} (1-\gamma B)\, a_{p+k} - ((1-\gamma A)\, p + (A-B)\beta\gamma)\, \frac{p^\lambda (p-\alpha)}{(p+k-\alpha)} \le (A-B)\gamma p^\lambda (p-\beta). \tag{2.2}$$

Proof. Since $f \in T_p^*(g(z), \alpha, \beta, \gamma, A, B, \lambda)$, there exists a function $g(z) \in T_p^*(\alpha, \lambda)$ such that

$$\left| \frac{z(D^\lambda f(z))' - pD^\lambda g(z)}{(A-B)(p-\beta)D^\lambda g(z) - B\left[z(D^\lambda f(z))' - pD^\lambda g(z)\right]} \right| < \gamma, \quad z \in U.$$

Then, we have

$$\begin{aligned} &\left| \frac{z(D^\lambda f(z))' - pD^\lambda g(z)}{(A-B)(p-\beta)D^\lambda g(z) - B\left[z(D^\lambda f(z))' - pD^\lambda g(z)\right]} \right| \\ &= \left| \sum_{k=1}^{\infty} \left((p+k)^{\lambda+1} a_{p+k} - p(p+k)^\lambda b_{p+k}\right) z^{p+k} \Big/ \left\{ (A-B)(p-\beta)p^\lambda z^p \right.\right. \\ &\left.\left. + \sum_{k=1}^{\infty} \left((p+k)^{\lambda+1} B a_{p+k} - \left[(A-B)(p-\beta)(p+k)^\lambda + Bp(p+k)^\lambda\right] b_{p+k}\right) z^{p+k} \right\} \right| \\ &< \gamma. \end{aligned}$$

Since $\mathrm{Re}(z) \leq |z|$ for all z, we obtain

$$\begin{aligned} \mathrm{Re}&\left[\sum_{k=1}^{\infty} \left((p+k)^{\lambda+1} a_{p+k} - p(p+k)^\lambda b_{p+k}\right) z^{p+k} \Big/ \left\{ (A-B)(p-\beta)p^\lambda z^p \right.\right. \\ &\left.\left. + \sum_{k=1}^{\infty} \left((p+k)^{\lambda+1} B a_{p+k} - \left[(A-B)(p-\beta)(p+k)^\lambda + Bp(p+k)^\lambda\right] b_{p+k}\right) z^{p+k} \right\} \right] \\ &< \gamma. \end{aligned} \tag{2.3}$$

We choose values of z in the real axis so that $\frac{z(D^\lambda f(z))'}{D^\lambda g(z)}$ is real. Then, upon simplifying the denominator in (2.3) and letting $z \to 1^-$ through real values, we obtain

$$\begin{aligned} &\sum_{k=1}^{\infty} (p+k)^{\lambda+1} a_{p+k} - p(p+k)^\lambda b_{p+k} \\ &\qquad \leq (A-B)\gamma p^\lambda (p-\beta) + \gamma \sum_{k=1}^{\infty} (p+k)^{\lambda+1} B a_{p+k} \\ &\qquad\qquad - \left[(A-B)(p-\beta)(p+k)^\lambda + Bp(p+k)^\lambda\right] b_{p+k} \end{aligned}$$

or, equivalently,

$$\begin{aligned} &\sum_{k=1}^{\infty} (p+k)^{\lambda+1} (1-\gamma B)\, a_{p+k} - ((1-\gamma A)\, p + (A-B)\beta\gamma)\, (p+k)^\lambda b_{p+k} \\ &\qquad \leq (A-B)\gamma p^\lambda (p-\beta). \end{aligned} \tag{2.4}$$

But, in view of Lemma 2.1, $T_p^*(\alpha, \lambda)$ implies $b_{p+k} \leq \frac{p^\lambda (p-\alpha)}{(p+k-\alpha)}$, $k \geq 1$.
Making use of the above inequality in (2.4), we find the required result.

The result is sharp, the extremal function being

$$f(z)=z^{p}-\frac{p^{\lambda+1}(1-\gamma B)(p-\alpha)+(A-B)\gamma k p^{\lambda}(p-\beta)}{(p+k)^{\lambda+1}(1-\gamma B)(p+k-\alpha)}z^{p+k}$$

for $k\geq 1$, with respect to

$$g(z)=z^{p}-\frac{p^{\lambda}(p-\alpha)}{(p+k)^{\lambda}(p+k-\alpha)}z^{p+k}.$$

□

Corollary 2.3. *Let the function defined by (1.2) be in the class* $T_p^*(g(z),\alpha,\beta,\gamma,A,B,\lambda)$. *Then*

$$a_{p+k}\leq\frac{p^{\lambda+1}(1-\gamma B)(p-\alpha)+(A-B)\gamma k p^{\lambda}(p-\beta)}{(p+k)^{\lambda+1}(1-\gamma B)(p+k-\alpha)}$$

for $k\in\mathbb{N}$.

3. Distortion Theorems

Theorem 3.1. *If* $f\in T_p^*(g(z),\alpha,\beta,\gamma,A,B,\lambda)$, *then for* $|z|=r$,

$$r^{p}-\frac{p^{\lambda+1}(1-\gamma B)(p-\alpha)+(A-B)\gamma p^{\lambda}(p-\beta)}{(p+1)^{\lambda+1}(1-\gamma B)(p+1-\alpha)}r^{p+1}\leq|f(z)|\leq r^{p}+\frac{p^{\lambda+1}(1-\gamma B)(p-\alpha)+(A-B)\gamma p^{\lambda}(p-\beta)}{(p+1)^{\lambda+1}(1-\gamma B)(p+1-\alpha)}r^{p+1}\qquad(3.1)$$

and

$$pr^{p-1}-\frac{p^{\lambda+1}(1-\gamma B)(p-\alpha)+(A-B)\gamma p^{\lambda}(p-\beta)}{(p+1)^{\lambda}(1-\gamma B)(p+1-\alpha)}r^{p}\leq\left|f'(z)\right|\leq pr^{p-1}+\frac{p^{\lambda+1}(1-\gamma B)(p-\alpha)+(A-B)\gamma p^{\lambda}(p-\beta)}{(p+1)^{\lambda}(1-\gamma B)(p+1-\alpha)}r^{p}.\qquad(3.2)$$

Proof. In view of (2.4), we have

$$(p+1)^{\lambda+1}(1-\gamma B)\sum_{k=1}^{\infty}a_{p+k}-(p+1)^{\lambda}\left((1-\gamma A)p+(A-B)\beta\gamma\right)\sum_{k=1}^{\infty}b_{p+k}\leq(A-B)\gamma p^{\lambda}(p-\beta).\qquad(3.3)$$

Again for $g(z)\in T_p^*(\alpha,\lambda)$, Lemma 2.1 yields

$$b_{p+k}\leq\frac{p^{\lambda}(p-\alpha)}{(p+k-\alpha)},\quad k\geq 1\qquad(3.4)$$

so that (3.3) reduces to

$$(p+1)^{\lambda+1}(1-\gamma B)\sum_{k=1}^{\infty}a_{p+k}-(p+1)^{\lambda}\left((1-\gamma A)p+(A-B)\beta\gamma\right)\frac{p^{\lambda}(p-\alpha)}{(p+k-\alpha)} \leq (A-B)\gamma p^{\lambda}(p-\beta).$$

Consequently,

$$\begin{aligned}|f(z)| &\geq |z|^{p}-\sum_{k=1}^{\infty}a_{p+k}|z|^{p+k}\\ &\geq |z|^{p}-|z|^{p+1}\sum_{k=1}^{\infty}a_{p+k}\\ &\geq |z|^{p}-\frac{p^{\lambda+1}(1-\gamma B)(p-\alpha)+(A-B)\gamma p^{\lambda}(p-\beta)}{(p+1)^{\lambda+1}(1-\gamma B)(p+1-\alpha)}|z|^{p+1}\end{aligned}$$

and

$$\begin{aligned}|f(z)| &\leq |z|^{p}+\sum_{k=1}^{\infty}a_{p+k}|z|^{p+k}\\ &\leq |z|^{p}+|z|^{p+1}\sum_{k=1}^{\infty}a_{p+k}\\ &\leq |z|^{p}+\frac{p^{\lambda+1}(1-\gamma B)(p-\alpha)+(A-B)\gamma p^{\lambda}(p-\beta)}{(p+1)^{\lambda+1}(1-\gamma B)(p+1-\alpha)}|z|^{p+1}.\end{aligned}$$

Furthermore, we have from (2.4) that

$$(p+1)^{\lambda}(1-\gamma B)\sum_{k=1}^{\infty}(p+k)a_{p+k}-(p+1)^{\lambda}\left((1-\gamma A)p+(A-B)\beta\gamma\right)\sum_{k=1}^{\infty}b_{p+k} \leq (A-B)\gamma p^{\lambda}(p-\beta),$$

which on applying (3.4) gives

$$\sum_{k=1}^{\infty}(p+k)a_{p+k}\leq\frac{p^{\lambda}\left((1-\gamma B)p(p-\alpha)+(A-B)\gamma(p-\beta)\right)}{(p+1)^{\lambda}(1-\gamma B)(p+1-\alpha)}.$$

Thus, we find

$$\begin{aligned}\left|f'(z)\right| &\geq p|z|^{p-1}-\sum_{k=1}^{\infty}(p+k)a_{p+k}|z|^{p+k-1}\\ &\geq p|z|^{p-1}-|z|^{p}\sum_{k=1}^{\infty}(p+k)a_{p+k}\\ &\geq p|z|^{p-1}-\frac{p^{\lambda}\left((1-\gamma B)p(p-\alpha)+(A-B)\gamma(p-\beta)\right)}{(p+1)^{\lambda}(1-\gamma B)(p+1-\alpha)}|z|^{p}\end{aligned}$$

and

$$\begin{aligned}
|f'(z)| &\le p|z|^{p-1} + \sum_{k=1}^{\infty}(p+k)a_{p+k}|z|^{p+k-1} \\
&\le p|z|^{p-1} + |z|^{p}\sum_{k=1}^{\infty}(p+k)a_{p+k} \\
&\le p|z|^{p-1} + \frac{p^{\lambda}\left((1-\gamma B)\,p(p-\alpha)+(A-B)\gamma(p-\beta)\right)}{(p+1)^{\lambda}(1-\gamma B)\,(p+1-\alpha)}|z|^{p},
\end{aligned}$$

which completes the proof of the theorem. □

The above bounds are sharp for the function $f(z)$ defined by

$$f(z) = z^{p} - \frac{p^{\lambda+1}(1-\gamma B)\,(p-\alpha)+(A-B)\gamma p^{\lambda}(p-\beta)}{(p+1)^{\lambda+1}(1-\gamma B)\,(p+1-\alpha)}z^{p+1} \tag{3.5}$$

with respect to

$$g(z) = z^{p} - \frac{p^{\lambda}(p-\alpha)}{(p+1)^{\lambda}(p+1-\alpha)}z^{p+1}.$$

Corollary 3.2. *Let $f \in T_p^*(g(z),\alpha,\beta,\gamma,A,B,\lambda)$.Then the unit disk U is mapped onto a domain that contains the disk*

$$|w| < \frac{\{[(p+1)^{\lambda+1}-p^{\lambda+1}](p-\alpha)+(p+1)^{\lambda+1}\}\,(1-\gamma B)-(A-B)\gamma p^{\lambda}(p-\beta)}{(p+1)^{\lambda+1}(1-\gamma B)\,(p+1-\alpha)}.$$

The result is sharp with the extremal function defined by (3.5).

The result follows upon letting $z \to 1^{-}$ in the left hand inequality of (3.1).

4. Extreme Points for $T_p^*(g(z),\alpha,\beta,\gamma,A,B,\lambda)$

Theorem 4.1. *Let $f_p(z) = z^p$ and*

$$f_{p+k}(z) = z^{p} - \frac{p^{\lambda+1}(1-\gamma B)\,(p-\alpha)+(A-B)\gamma k p^{\lambda}(p-\beta)}{(p+k)^{\lambda+1}(1-\gamma B)\,(p+k-\alpha)}z^{p+k}.$$

If $f \in T_p^*(g(z),\alpha,\beta,\gamma,A,B,\lambda)$ then it can be expressed in the form

$$f(z) = \sum_{k=0}^{\infty}\xi_{p+k}f_{p+k}(z),$$

where $\xi_{p+k} \ge 0$ and $\sum_{k=0}^{\infty}\xi_{p+k} = 1$.

Proof. Suppose that $f \in T_p^*(g(z),\alpha,\beta,\gamma,A,B,\lambda)$. Since

$$a_{p+k} \le \frac{p^{\lambda+1}(1-\gamma B)\,(p-\alpha)+(A-B)\gamma k p^{\lambda}(p-\beta)}{(p+k)^{\lambda+1}(1-\gamma B)\,(p+k-\alpha)},$$

we may set

$$\xi_{p+k} = \frac{(p+k)^{\lambda+1}(1-\gamma B)(p+k-\alpha)}{p^{\lambda+1}(1-\gamma B)(p-\alpha)+(A-B)\gamma k p^{\lambda}(p-\beta)} a_{p+k}$$

and $\xi_p = 1 - \sum_{k=1}^{\infty} \xi_{p+k}$. Then

$$\begin{aligned}
f(z) &= z^p - \sum_{k=1}^{\infty} a_{p+k} z^{p+k} \\
&= z^p - \sum_{k=0}^{\infty} \xi_{p+k} \frac{p^{\lambda+1}(1-\gamma B)(p-\alpha)+(A-B)\gamma k p^{\lambda}(p-\beta)}{(p+k)^{\lambda+1}(1-\gamma B)(p+k-\alpha)} z^{p+k} \\
&= z^p - \sum_{k=1}^{\infty} \xi_{p+k}(z^p - f_{p+k}(z)) \\
&= z^p - \sum_{k=1}^{\infty} \xi_{p+k} z^p - \sum_{k=1}^{\infty} \xi_{p+k} f_{p+k}(z) \\
&= \left(1 - \sum_{k=1}^{\infty} \xi_{p+k}\right) z^p + \sum_{k=1}^{\infty} \xi_{p+k} f_{p+k}(z) \\
&= \sum_{k=0}^{\infty} \xi_{p+k} z^{p+k}.
\end{aligned}$$

This completes the proof of theorem. □

Bibliography

[1] J. Patel and S. Rout, On a certain subclasses of $p-$valent starlike functions, *Kyungpook Math. J.*, **36** (1996), 13–27.

[2] H.M. Srivastava and S. Owa, On certain subclasses of starlike functions, *J. Math. Anal. Appl.*, **161** (1991), 405–415.

[3] V.P. Gupta and P.K. Jain, Certain classes of univalent functions with negative coefficients II, *Bull. Austral. Math. Soc.*, **15** (1976), 467-473.

[4] S.M. Sarangi and B.A. Uralegaddi, The radius of convexity and starlikeness for certain classes of analytic functions with negative coefficients I, *Lincei. Rend. Sc. fis. mat. e nat.*, **LXV** (1978), 38–42.

[5] G.S. Salagean, Subclasses of univalent functions, *Lecture notes in Mathematics*, 1013-Springer Verlag, Berlin, Heidelberg and New York, 1983, 362-372.

In: Advances in Inequalities for Series
Editors: S. S. Dragomir and A. Sofo, pp. 131–141
ISBN 978-1-60021-920-7

Chapter 7

A New Sub-class of Multivalent Functions

S. Latha[1] *and O. Karthiyayini*[2]
[1]Department of Mathematics, Maharaja's College
University of Mysore, Mysore - 570005, INDIA
E-mail address: drlatha@gmail.com
[2]Department of Mathematics
Vidya Vikas Institute of Engineering and Technology
Mysore - 570011, INDIA
E-mail address: karthiyayino@yahoo.co.in

Abstract

A class of multivalent functions is introduced by using the Ruscheweyh derivative and the Hadamard product. The object of the present paper is to obtain the coefficient estimates, extreme points, closure theorems and distortion theorems of functions in this class.

1991 Mathematics Subject Classification: 30C45

Key words and phrases: Multivalent, Hadamard product, Ruscheweyh derivatives, Pochhammer symbol.

1. Introduction

Let T_p denote the family of functions of the form,

$$f(z) = z^p + \sum_{m=p+1}^{\infty} a_m z^m \tag{1.1}$$

which are analytic and multivalent in the open disc $U = \{z : |z| < 1\}$.

The class T_p is closed under the Hadamard product or convolution. The convolution of two functions $f(z), g(z) \in T_p$ is defined by,

$$(f * g)(z) = z^p + \sum_{m=p+1}^{\infty} a_m b_m z^m \qquad p \in \mathbb{N} = \{1, 2, 3, ...\} \tag{1.2}$$

where,

$$f(z)=z^p+\sum_{m=p+1}^{\infty} a_m z^m;\quad g(z)=z^p+\sum_{m=p+1}^{\infty} b_m z^m$$

We define the function $\phi_p(a,c;z)$ by

$$\phi_p(a,c;z)=z^p+\sum_{m=p+1}^{\infty}\frac{(a)_m}{(c)_m}z^m \tag{1.3}$$

for $c\neq 0,-1,-2,\ldots,a\neq -1,\ z\in U$, where $(\lambda)_m$ is the Pochhammer symbol defined by,

$$(\lambda)_m=\frac{\Gamma(m+\lambda)}{\Gamma(\lambda)}$$

$$\text{i.e.,}\quad (\lambda)_0=1,(\lambda)_m=m(m+1)\cdots(\lambda+m-1),\quad m>1. \tag{1.4}$$

Consider a linear operator $L_p(a,c)$ defined by,

$$L_p(a,c)f(z)=\phi_p(a,c;z)*f(z)=z^p+\sum_{m=p+1}^{\infty}\frac{(a)_m}{(c)_m}a_m z^m. \tag{1.5}$$

We note that $L_p(a,a)f(z)=f(z),\ L_p(2,1)f(z)=zf'(z)$ and

$$L_p(n+1,1)f(z)=D^n f(z),$$

where $D^n f(z)$ is the Ruscheweyh derivative of $f(z)$ defined by,

$$D^n f(z)=\frac{z}{(1-z)^{n+1}}*f(z),\qquad n>-1. \tag{1.6}$$

Equivalently, we have,

$$D^n f(z)=\frac{z(z^{n-1}f(z))^{(n)}}{n!}.$$

Therefore,

$$L_p(n+1,1)f(z)=\frac{\Gamma(n+p)z^p}{(p-1)!\Gamma(n+1)}+\sum_{m=p+1}^{\infty}\frac{\Gamma(n+m)a_m z^m}{(m-1)!\Gamma(n+1)}. \tag{1.7}$$

Let Ω denote the family of function ω which are analytic in U and satisfy the condition $\omega(0)=0$ and $|\omega(z)|<1$ for $z\in U$. For two real numbers $A,B\quad(-1\leq B<A\leq 1)$, $P(A,B)$ denotes the class of analytic functions P defined on U with the representations,

$$P(z)=\frac{1+A\omega(z)}{1+B\omega(z)},\quad Z\in U\quad\text{and}\quad \omega\in\Omega. \tag{1.8}$$

Also, let $P(A,B,p,\alpha)$ be the class of functions of T_p which are of the form,

$$p(z)=p+\sum_{m=1}^{\infty}a_m z^m \tag{1.9}$$

analytic in U and

$$p(z) = (p-\alpha)p_1(z) + \alpha,$$

where $0 \le \alpha \le p$, $p_1(z) \in P(A,B)$. Also, every $p(z) \in P(A,B,p,\alpha)$ has a representation of the form,

$$f(z) = \frac{p+\gamma\omega(z)}{1+B\omega(z)}, \tag{1.10}$$

where $\gamma = (p-\alpha)A + \alpha B$, $-1 \le B < A \le 1$, $0 \le \alpha < p$ and $\omega(z) \in \Omega$.

$$\text{For} \quad \phi(z) = z^p + \sum_{m=p+1}^{\infty} \lambda_m z^m \quad \text{and} \quad \psi(z) = z^p + \sum_{m=p+1}^{\infty} \mu_m z^m$$

which are analytic and multivalent in U such that $\lambda_m \ge 0$, $\mu_m \ge 0$ and $\lambda_m \ge \mu_m$, $m \ge p+1$, we let $E_n^p(\phi,\psi,A,B,\alpha)$ denote the class of functions $f(z) \in T_p$ which satisfy the analytic criterion, $(f*\psi)(z) \ne 0$ and

$$\frac{L_p(n+2,1)(f*\phi)(z)}{L_p(n+1,1)(f*\psi)(z)} \in P(A,B,p,\alpha), \tag{1.11}$$

where, $-1 \le B < A \le 1$, $0 \le \alpha < P$, $p \in \mathbb{N} = \{1,2,3,\ldots\}$.

Let T denote the sub-class of T_p consisting of functions analytic and multivalent in U which can be expressed in the form,

$$f(z) = z^p - \sum_{m=p+1}^{\infty} a_m z^m, \quad a_m \ge 0. \tag{1.12}$$

Let us define $E_n^p[\phi,\psi,A,B,\alpha] = E_n^p(\phi,\psi,A,B,\alpha) \cap T$. For specific values of ϕ,ψ,A,B,p,α we obtain the following classes studied by various authors in literature.

$$E_n^1[\phi,\psi,A,B,0] = E_n[\phi,\psi,A,B],$$

$$E_n^1\left[\frac{z}{(1-z)}, z, A, B, 0\right] = V_n[A,B],$$

$$E_n^1\left[\frac{z}{1-z}, \frac{z}{1-z}, A, B, 0\right] = S_n[A,B],$$

$$E_n^1\left[\frac{z}{(1-z)^2}, \frac{z}{(1-z)^2}, A, B, 0\right] = K_n[A,B],$$

$$E_0^1\left[\frac{z}{1-z^{2(1-\gamma)}}, z, (2\alpha-1)\beta, \beta, 0\right] = p_\gamma[\alpha,\beta]$$

$$0 \le \alpha < 1, 0 < \beta \le 1, 0 \le \gamma < 1,$$

$$E_0^1\left[\frac{z}{(1-z)^{2(1-\gamma)}}, \frac{z}{(1-z)^{2(1-\gamma)}}, (2\alpha-1)\beta, \beta, 0\right] = R_\gamma[\alpha,\beta].$$

The class $E_n[\phi,\psi,A,B]$ was introduced and studied by M.L. Mogra [5]. The class $V_n[A,B]$ was studied by Kumar [3]. Padmanabhan and Manjini [7] introduced and studied the classes $S_n[A,B]$ and $K_n[A,B]$. The class $P_\gamma[A,B]$ was studied by Owa and Ahuja [6] and the class $R_\gamma[A,B]$ was studied by Ahuja and Silverman [1]. Several other classes introduced and studied by Silverman [9], Silverman and Silvia [10], Gupta and Jain [2] and others can be obtained from the class $E_n^p[\phi,\psi,A,B,\alpha]$.

The aim of the present paper is to obtain necessary and sufficient conditions for a function $f(z)$ to be in $E_n^p[\phi,\psi,A,B,\alpha]$. Furthermore, extreme points, closure properties and distortion bounds for $f(z)\in E_n^p[\phi,\psi,A,B,\alpha]$ are also obtained.

2. Coefficient Inequalities

Theorem 2.1. *A sufficient condition for a function $f(z)$ of the form (1.1) to be in $E_n^p(\phi,\psi,A,B,\alpha)$ is that,*

$$\sum_{m=p+1}^{\infty}\frac{(n+m-1)!\sigma_m}{(m-1)!(n+1)!}|a_m|\le\frac{(n+p-1)!\tau_p}{(p-1)!(n+1)!},\tag{2.1}$$

where,

$$\begin{aligned}&-1\le A<B\le 1,\quad 0\le\alpha<p,\quad p\in\mathbb{N}=\{1,2,3,\dots\},\\&\sigma_m=(B+1)(n+m)\lambda_m-[p(A+1)-\alpha(A-B)](n+1)\mu_m,\\&\tau_p=B(n+p)-(n+1)[(p-\alpha)A+\alpha B]+n(p-1).\end{aligned}$$

Proof. It suffices to show that,

$$\left|\frac{L(n+2,1)(f*\phi)(z)-pL(n+1,1)(f*\psi)(z)}{BL(n+2,1)(f*\phi)(z)-\gamma L(n+1,1)(f*\psi)(z)}\right|<1.\tag{2.2}$$

For $|z|=r,\quad 0\le r<1$ let us consider,

$$\begin{aligned}&|L(n+2,1)(f*\phi)(z)-pL(n+1,1)(f*\psi)(z)|\\&\quad-|BL(n+2,1)(f*\phi)(z)-\gamma L(n+1,1)(f*\psi)(z)|\\&<\sum_{m=p+1}^{\infty}\frac{(n+m-1)!}{(m-1)!(n+1)!}\{(B+1)(n+m)\lambda_m-(p+\gamma)(n+1)\mu_m\}|a_m|r^m\\&\quad-\frac{(n+p-1)!}{(p-1)!(n+1)!}\{B(n+p)-\gamma(n+1)+n(p-1)\}r^p\\&<\left[\sum_{m=p+1}^{\infty}\frac{(n+m-1)!}{(m-1)!(n+1)!}\{(B+1)(n+m)\lambda_m-[p(A+1)-\alpha(A-B)](n+1)\mu_m\}|a_m|\right.\\&\quad\left.-\frac{(n+p-1)!}{(p-1)!(n+1)!}\{B(n+p)-(n+1)[(p-\alpha)A+\alpha B]+n(p-1)\}\right]|z|\\&\le 0,\end{aligned}$$

which implies that $f(z)\in E_n^p(\phi,\psi,A,B,\alpha)$. □

Corollary 2.2. *For $p=1,\quad \alpha=0$ we obtain Theorem 2.1 of [5].*

Theorem 2.3. *Any function $f(z)$ of the form $f(z)=z^p-\sum_{m=p+1}^{\infty}a_mz^m$ is in $E_n^p[\phi,\psi,A,B,\alpha]$ if and only if (2.1) is satisfied.*

Proof. In view of Theorem 2.1 we need to prove the necessity part only. If $f \in E_n^p[\phi,\psi,A,B,\alpha]$ then from (1.10) and (1.11) we have,

$$|w(z)| = \left| \left(\sum_{m=p+1}^{\infty} \frac{(n+m-1)!}{(m-1)!(n+1)!}[(n+m)\lambda_m - p(n+1)\mu_m]a_m z^{m-p} \right) \Big/ \left(\frac{(n+p-1)!\tau_p}{(p-1)!(n+1)!} + \sum_{m=p+1}^{\infty} \frac{(n+m-1)!}{(m-1)!(n+1)!} \times \{B(n+m)\lambda_m - \gamma(n+1)\mu_m\}a_m z^{m-p} \right) \right| < 1$$

where $\tau_p = B(n+p) - (n+1)[(p-\alpha)A - \alpha B] + n(p-1)$.

Hence,

$$\Re \left\{ \frac{\sum_{m=p+1}^{\infty} \frac{(n+m-1)!}{(m-1)!(n+1)!}[(n+m)\lambda_m - p(n+1)\mu_m]a_m z^{m-p}}{\frac{(n+p-1)!\tau_p}{(p-1)!(n+1)!} + \sum_{m=p+1}^{\infty} \frac{(n+m-1)!}{(m-1)!(n+1)!}\{B(n+m)\lambda_m - \gamma(n+1)\mu_m\}a_m z^{m-p}} \right\} < 1, \quad (2.3)$$

where $\tau_p = B(n+p) - (n+1)[(p-\alpha)A - \alpha B] + n(p-1)$ for all $z \in U$.

Taking real values of z, $z = r$, $0 < r < 1$ in (2.3) and letting $r \to 1$ we obtain the inequality (2.1). □

Corollary 2.4. *For $p = 1$, $\alpha = 0$ we obtain Theorem 2.2 of [5].*

Corollary 2.5. *If $f(z) \in E_n^p[\phi,\psi,A,B,\alpha]$ then*

$$a_m \le \frac{(n+p-1)!(m-1)!\tau_p}{(n+m-1)!(p-1)!\sigma_m} z^m, \qquad z \in U. \tag{2.4}$$

The equality holds for each m, for functions of the form,

$$f_m(z) = z^p - \frac{(n+p-1)!(m-1)!\tau_p}{(n+m-1)!(p-1)!\sigma_m} z^m, \qquad z \in U. \tag{2.5}$$

Corollary 2.6. *Putting $p = 1$ and $\alpha = 0$ in Corollary 2.5, we obtain Corollary 2.1 of [5].*

3. Closure Theorems

Theorem 3.1. *The class $E_n^p[\phi,\psi,A,B,\alpha]$ is closed under convex linear combinations.*

Proof. Let

$$f(z) = z^p - \sum_{m=p+1}^{\infty} a_m z^m \qquad \text{and} \qquad g(z) = z^p - \sum_{m=p+1}^{\infty} b_m z^m,$$

where $a_m \geq 0,\ b_m \geq 0$. Assuming that $f(z), g(z) \in E_n^p[\phi, \psi, A, B, \alpha]$, it suffices to show that the function $h(z)$ defined by,

$$h(z) = \eta f(z) + (1-\eta)g(z) \qquad (0 \leq \eta \leq 1) \tag{3.1}$$

is also in $E_n^p[\phi, \psi, A, B, \alpha]$.

Since,

$$h(z) = z^p - \sum_{m=p+1}^{\infty} \frac{(m+n-1)!\sigma_m}{(m-1)!(n+1)!}[\eta a_m + (1-\eta)b_m],$$

by virtue of Theorem 2.3 we have,

$$\sum_{m=p+1}^{\infty} \frac{(n+m-1)!\sigma_m}{(m-1)!(n+1)!}[\eta a_m + (1-\eta)b_m]$$
$$\leq \eta\left[\frac{(n+p-1)!\tau_p}{(p-1)!(n+1)!}\right] + (1-\eta)\left[\frac{(n+p-1)!\tau_p}{(p-1)!(n+1)!}\right] = \frac{(n+p-1)!\tau_p}{(p-1)!(n+1)!},$$

which implies $h(z) \in E_n^p[\phi, \psi, A, B, \alpha]$. □

Theorem 3.2. *Let the functions*

$$f_j(z) = z^p - \sum_{m=p+1}^{\infty} |a_{nj}| z^m, \quad a_{nj} \geq 0, \quad j = 1, 2, 3, \ldots, k \tag{3.2}$$

be in the class $E_n^p[\phi, \psi, A, B, \alpha]$. *Then the function,*

$$h(z) = z^p - \sum_{m=p+1}^{\infty} |c_m| z^m \tag{3.3}$$

is also in $E_n^p[\phi, \psi, A, B, \alpha]$, *where*

$$c_m = \frac{1}{k}\sum_{j=1}^{k} a_{nj}. \tag{3.4}$$

Proof. From (3.2) and (3.3) we have,

$$h(z) = z^p - \sum_{m=p+1}^{\infty}\left[\frac{1}{k}\sum_{j=1}^{k}|a_{nj}|\right] z^m.$$

Since $f_j \in E_n^p[\phi, \psi, A, B, \alpha]$ from Theorem 2.3 we have,

$$\sum_{m=p+1}^{\infty} \frac{(n+m-1)!\sigma_m}{(m-1)!(n+1)!}|a_{nj}| \leq \frac{(n+p-1)!\tau_p}{(p-1)!(n+1)!}. \tag{3.5}$$

Using (3.4) and (3.5) we have,

$$\sum_{m=p+1}^{\infty} \frac{(n+m-1)!\sigma_m}{(m-1)!(n+1)!}|c_m| = \sum_{m=p+1}^{\infty} \frac{(n+m-1)!\sigma_m}{(m-1)!(n+1)!}\left[\frac{1}{k}\sum_{j=1}^{k} a_{nj}\right] \leq \frac{(n+p-1)!\tau_p}{(p-1)!(n+1)!}$$

in view of Theorem 2.3. Hence $h(z) \in E_n^p[\phi, \psi, A, B, \alpha]$. □

Theorem 3.3. *Let* $f_p(z) = z^p$,

$$f_m(z) = z^p - \frac{(n+p-1)!(m-1)!\tau_p}{(n+m-1)!(p-1)!\sigma_m} z^m \quad (m \geq p+1).$$

Then $f(z)$ *is in* $E_n^p[\phi, \psi, A, B, \alpha]$ *if and only if it can be expressed as,*

$$f(z) = \sum_{m=p}^{\infty} \rho_m f_m(z), \tag{3.6}$$

where $\rho_m \geq 0$ *and* $\sum_{m=p}^{\infty} \rho_m = 1$

Proof. Let us suppose that $f(z)$ can be expressed in the form (3.6). Then we have,

$$f(z) = \sum_{m=p}^{\infty} \rho_m f_m(z) = z^p - \sum_{m=p+1}^{\infty} \frac{(n+p-1)!(m-1)!\tau_p}{(n+m-1)!(p-1)!\sigma_m} \rho_m z^m.$$

Therefore,

$$\sum_{m=p+1}^{\infty} \left[\frac{(n+m-1)!\sigma_m}{(m-1)!(n+1)!} \times \frac{(n+p-1)!(m-1)!\tau_p}{(n+m-1)!(p-1)!\sigma_m} \right] \rho_m = \sum_{m=p+1}^{\infty} \frac{(n+p-1)!\tau_p}{(p-1)!(n+1)!} \rho_m$$
$$\leq \frac{(n+p-1)!\tau_p}{(p-1)!(n+1)!}.$$

Hence it follows from Theorem 2.3 that $f(z)$ is in $E_n^p[\phi, \psi, A, B, \alpha]$.

Conversely, let us suppose that $f(z)$ defined by (1.12) is in the class $E[\phi, \psi, A, B, \alpha]$. Then using (2.4) we have,

$$|a_m| \leq \frac{(n+p-1)!(m-1)!\tau_p}{(n+m-1)!(p-1)!\sigma_m}.$$

Setting,

$$\rho_m = \frac{(n+m-1)!(p-1)!\sigma_m}{(n+p-1)!(m-1)!\tau_p}$$

and

$$\rho_p = 1 - \sum_{m=p+1}^{\infty} \rho_m,$$

Theorem 2.3 yields, $\sum_{m=p+1}^{\infty} \rho_m \leq 1$ and hence $\rho_p \geq 0$. This completes the proof. □

Corollary 3.4. *For* $p = 1$, $\alpha = 0$ *we obtain Theorem 3.2 of [5].*

Corollary 3.5. *The extreme points of* $E_n^p[\phi, \psi, A, B, \alpha]$ *are the functions* $f_m(z)$, $m \geq p+1$, $p \in \mathbb{N} = \{1, 2, 3, \ldots\}$.

Corollary 3.6. *For* $p = 1$, $\alpha = 0$ *in Corollary 3.5 we obtain Corollary 3.1 of [5]*

4. Distortion Theorems

Theorem 4.1. *Let* $f \in E_n^p[\phi, \psi, A, B, \alpha]$ *and let*

$$\sigma_m = (B+1)(n+m)\lambda_m - [p(A+1) - \alpha(A-B)](n+1)\mu_m,$$

$$\tau_p = B(n+p) - (n+1)[(p-\alpha)A + \alpha B] + n(p-1), \quad m \geq p+1.$$

If $n, m, \sigma_m, \sigma_{m+1}$ *and* $|z|$ *satisfy the condition,*

$$(n+m)\sigma_{m+1} - m\sigma_m|z| \geq 0, \tag{4.1}$$

then

$$\max\left\{0, |z|^p - \frac{(n+p-1)!\tau_p}{(p-1)!(n+1)!\sigma_{p+1}}|z|^{p+1}\right\} \leq |f(z)| \leq |z|^p + \frac{(n+p-1)!\tau_p}{(p-1)!(n+1)!\sigma_{p+1}}|z|^{p+1}. \tag{4.2}$$

The bounds are sharp.

Proof. By virtue of Theorem 3.3 we have,

$$|f(z)| \geq \max\left\{0, |z|^p - \max_m \frac{(n+p-1)!(m-1)!\tau_p}{(n+m-1)!(p-1)!\sigma_m}|z|^m\right\} \tag{4.3}$$

and

$$|f(z)| \leq z^p + \max_m \frac{(n+p-1)!(m-1)!\tau_p}{(p-1)!(n+m-1)!\sigma_m}|z|^m \tag{4.4}$$

for $z \in U$. It suffices to show that,

$$J(A, B, n, m, \sigma_m, \tau_p, |z|) = \frac{(n+p-1)!(m-1)!\tau_p}{(n+m-1)!(p-1)!\sigma_m}|z|^m$$

is a decreasing function of m $(m \geq p+1)$. For $|z| \neq 0$, it can be observed that,

$$J(A, B, n, m, \sigma_m, \tau_p, |z|) \geq J(A, B, n, m+1, \sigma_{m+1}, \tau_p, |z|)$$

if and only if,

$$(n+m)\sigma_{m+1} - m\sigma_m|z| \geq 0.$$

Therefore $\max_m J(A, B, n, m, \sigma_m, \tau_p, |z|)$ is attained at $m = p+1$ which implies (4.2). □

Corollary 4.2. *For* $p = 1$, $\alpha = 0$, *we obtain Theorem 4.1 of [5]*

Theorem 4.3. *Let* $f \in E_n^p[\phi, \psi, A, B, \alpha]$ *and let*

$$\sigma_m = (B+1)(n+m)\lambda_m - [p(A+1) - \alpha(A-B)](n+1)\mu_m$$
$$\tau_p = B(n+p) - (n+1)[(p-\alpha)A + \alpha B] + n(p-1), \quad m \geq p+1.$$

If $n, m, \sigma_m, \sigma_{m+1}$ *and* $|z|$ *satisfy the condition,*

$$(n+m)\sigma_{m+1} - (m+1)\sigma_m|z| \geq 0 \tag{4.5}$$

then

$$\max\left\{0, p|z|^{p-1} - \frac{(n+p-1)!(p+1)\tau_p}{(p-1)!(n+1)!\sigma_{p+1}}|z|^p\right\}$$
$$\leq |f'(z)| \leq p|z|^{p-1} + \frac{(n+p-1)!(p+1)\tau_p}{(p-1)!(n+1)!\sigma_{p+1}}|z|^p. \tag{4.6}$$

The bounds are sharp for the function

$$f(z) = z^p - \left[\frac{(p+1)(B-A)}{\sigma_{p+1}}\right] z^{p+1}.$$

Proof. From Theorem 3.3 it follows that,

$$|f'(z)| \geq p|z|^{p-1} - \max_m \frac{(n+p-1)!m!\tau_p}{(n+m-1)!(p-1)!\sigma_m}|z|^{m-1} \tag{4.7}$$

and

$$|f'(z)| \leq p|z|^{p-1} + \max_m \frac{(n+p-1)!m!\tau_p}{(n+m-1)!(p-1)!\sigma_m}|z|^{m-1}. \tag{4.8}$$

It suffices to show that,

$$J^*(A,B,m,n,\sigma_m,|z|) = \frac{(n+p-1)!m!\tau_p}{(n+m-1)!(p-1)!\sigma_m}|z|^{m-1}$$

is a decreasing function of m $(m \geq p+1)$. For $|z| \neq 0$, it can be easily checked that

$$J^*(A,B,m,n,\sigma_m,|z|) \geq J^*(A,B,m+1,n,\sigma_{m+1},|z|)$$

if and only if,

$$(n+m)\sigma_{m+1} - (m+1)\sigma_m|z| \geq 0.$$

Therefore, $\max_m J^\star(A,B,m,n,\sigma_m,|z|)$ is attained at $m = p+1$ and hence (4.6) follows.

□

Corollary 4.4. *If* $p = 1$, $\alpha = 0$ *then Theorem 4.2 of [5] is obtained.*

5. Integral Operators

Theorem 5.1. *Let* c *be a real number such that* $c > -p$. *If* $f(z) \in E_n^p[\phi,\psi,A,B,\alpha]$ *then the function* $F(z)$ *defined by,*

$$F(z) = \frac{c+p}{z^c}\int_0^z t^{c-1}f(t)dt \tag{5.1}$$

also belongs to $E_n^p[\phi,\psi,A,B,\alpha]$.

Proof. Let

$$f(z) = z^p - \sum_{m=p+1}^{\infty} |a_m| z^m.$$

Then (5.1) yields,

$$F(z) = z^p - \sum_{m=p+1}^{\infty} |b_m| z^m,$$

where,

$$|b_m| = \left(\frac{c+p}{c+m}\right) |a_m|.$$

Hence we have,

$$\sum_{m=p+1}^{\infty} \frac{(n+m-1)!\sigma_m}{(m-1)!(n+1)!} |b_m| = \sum_{m=p+1}^{\infty} \frac{(n+m-1)!\sigma_m}{(m-1)!(n+1)!} \left(\frac{c+p}{c+m}\right) |a_m|$$
$$\leq \sum_{m=p+1}^{\infty} \frac{(n+m-1)!\sigma_m}{(m-1)!(n+1)!} |a_m| \leq \frac{(n+p-1)!\tau_p}{(p-1)!(n+1)!}.$$

Therefore, from Theorem 2.3 it follows that $F(z) \in E_n^p[\phi, \psi, A, B, \alpha]$. □

Bibliography

[1] O.P. Ahuja and H. Silverman, Convolutions of prestarlike functions, *Internat. J. Math. and Math. Sci.,* **6**(1) (1983), 59–68.

[2] V.P. Gupta and P.K. Jain, Certain classes of univlent functions with negative coefficients, *II, Bull. Austral, Math. Soc.,* **15** (1976), 467–473.

[3] V. Kumar, On a new criterion for univalent functions, *Demonstratio Math.,* **17**(4) (1984), 875–886.

[4] S. Latha, Some studies in the theory of univalent and multivalent functions, *Ph.D Thesis,* (1994)

[5] M.L. Mogra, Applications of Ruscheweyh derivatives and Hadamard products to analytic functions, *Internat. J. Math and Math. Sci.,* **22**(4) (1999), 795–805.

[6] S. Owa and O.P. Ahuja, A class of functions defined by using Hadamard products, *Hokkaido Math. J.,* **15**(2) (1986), 217–232.

[7] K.S. Padmanabhan and R. Manjini, Certain classes of analytic functions with negative coefficients, *Indian J. Pure Appl. Math.,* **17**(10) (1986), 1210–1223.

[8] S. Ruscheweyh, New criteria for univalent functions, *Proc. Amer. Math. Soc.,* **49** (1975), 109–115.

[9] H. Silverman, Univalent functions with negative coefficients, *Proc. Amer. Math. Soc.,* **51** (1975), 109–116.

[10] H. Silverman and E. Silvia, Subclasses of prestarlike functions, *Math. Japan.,* **29**(6) (1984), 929–935.

In: Advances in Inequalities for Series
Editors: S. S. Dragomir and A. Sofo, pp. 143–149
ISBN 978-1-60021-920-7

Chapter 8

A Note on Coefficient Estimates for a Class of Analytic Functions

S. Latha[1] *and L. Shivarudrappa*[1]
[1]Department of Mathematics, Maharaja's College
University of Mysore, Mysore - 570005, INDIA.
E-mail address: drlatha@gmail.com
[2]Department of Mathematics, MMK and SDM Mahila Maha Vidyalaya
Mysore - 570004, INDIA
E-mail address: shivarudrappa@lycos.com

Abstract

The object of this paper is to obtain coefficient estimates for functions belonging to $M^{\delta}(b,\alpha)$ which generalizes some well-known classes studied in the literature.

1991 Mathematics Subject Classification: 30C45

Key words and phrases: Analytic functions, Univalent functions, Starlike functions, Convex functions, Hadamard product.

1. Introduction

Let N denote the family of analytic functions in the unit disc $U = \{z : |z| < 1\}$ normalized by the conditions $f(0) = 0 = f'(0) - 1$. Let $M(\alpha)$ be the subclass of N consisting of functions f which satisfy the inequality

$$\Re\left\{z\frac{f'(z)}{f(z)}\right\} < \alpha \qquad (z \in U) \tag{1.1}$$

for some $\alpha > 1$.

Let $N(\alpha)$ be the subclass of N consisting of functions f which obey the inequality

$$\Re\left\{z\frac{f''(z)}{f'(z)} + 1\right\} < \alpha \qquad (z \in U) \tag{1.2}$$

for some $\alpha > 1$ ([1] and [2]). Motivated by (1.1) and (1.2) we make the following definition.

Let $M^\delta(b,\alpha)$ be the class of functions $f \in N$ such that

$$\Re\left\{1-\frac{2}{b}+\frac{2}{b}\frac{D^{\delta+2}f(z)}{D^{\delta+1}f(z)}\right\} < \alpha, \qquad (z \in U), \tag{1.3}$$

where b is a non-zero complex number, $\alpha > 1$ and

$$D^\delta f(z) = \frac{z}{(1-z)^{\delta+1}} \star f(z), \quad \delta \geq -1, \tag{1.4}$$

where $\star$ denotes the convolution or the Hadamard product of the functions on the right. We note that

$$D^\delta f(z) = z + \sum_{n=2}^{\infty} \frac{\Gamma(n+\delta)}{(n-1)!\Gamma(1+\delta)} a_n z^n, \tag{1.5}$$

where

$$f(z) = z + \sum_{n=2}^{\infty} a_n z^n. \tag{1.6}$$

This class generalizes the classes studied by Owa and Nishiwaki ([1] and [2]) and Uralegaddi, Ganigi and Sarangi [4].

2. Coefficient Estimates

Our main goal in this section is to establish the following theorem.

Theorem 2.1. *If $f(z) \in N$ satisfies*

$$\sum_{n=2}^{\infty} \{[b(1-k)(\delta+2)+2(n-1)] + |b(1-2\alpha+k)(\delta+2)+2(n-1)|\} \frac{\Gamma(n+\delta+1)}{(n-1)!\Gamma(3+\delta)} |a_n| \leq 2b(\alpha-1), \tag{2.1}$$

where b is a non-zero complex number, $\delta \geq -1$, $0 \leq k \leq 1$ and $\alpha > 1$, then $f(z) \in M^\delta(b,\alpha)$.

Proof. Let us suppose (2.1). It suffices to show that

$$\left| \frac{1-\frac{2}{b}+\frac{2}{b}\frac{D^{\delta+2}f(z)}{D^{\delta+1}f(z)}-k}{1-\frac{2}{b}+\frac{2}{b}\frac{D^{\delta+2}f(z)}{D^{\delta+1}f(z)}-(2\alpha-k)} \right| < 1.$$

We note that

$$\left|\frac{1-\frac{2}{b}+\frac{2}{b}\frac{D^{\delta+2}f(z)}{D^{\delta+1}f(z)}-k}{1-\frac{2}{b}+\frac{2}{b}\frac{D^{\delta+2}f(z)}{D^{\delta+1}f(z)}-(2\alpha-k)}\right|$$

$$=\left|\frac{(b-2-kb)\left(z+\sum_{n=2}^{\infty}\frac{\Gamma(n+\delta+1)}{(n-1)!\Gamma(2+\delta)}a_nz^n\right)+2\left(z+\sum_{n=2}^{\infty}\frac{\Gamma(n+\delta+2)}{(n-1)!\Gamma(1+\delta)}a_nz^n\right)}{[b-2-(2\alpha-k)b]\left(z+\sum_{n=2}^{\infty}\frac{\Gamma(n+\delta+1)}{(n-1)!\Gamma(2+\delta)}a_nz^n\right)+2\left(z+\sum_{n=2}^{\infty}\frac{\Gamma(n+\delta+2)}{(n-1)!\Gamma(3+\delta)}a_nz^n\right)}\right|$$

$$=\left|\frac{b(1-k)+\sum_{n=2}^{\infty}[b(1-k)(\delta+2)+2(n-1)]\frac{\Gamma(n+\delta+1)}{(n-1)!\Gamma(3+\delta)}a_nz^{n-1}}{b(1-2\alpha+k)+\sum_{n=2}^{\infty}[b(1-2\alpha+k)(\delta+2)+2(n-1)]\frac{\Gamma(n+\delta+1)}{(n-1)!\Gamma(3+\delta)}a_nz^{n-1}}\right|$$

$$\leq\frac{b(1-k)+\sum_{n=2}^{\infty}[b(1-k)(\delta+2)+2(n-1)]\frac{\Gamma(n+\delta+1)}{(n-1)!\Gamma(3+\delta)}|a_n||z|^{n-1}}{b(1-2\alpha+k)-\sum_{n=2}^{\infty}|b(1-2\alpha+k)(\delta+2)+2(n-1)|\frac{\Gamma(n+\delta+1)}{(n-1)!\Gamma(3+\delta)}|a_n||z|^{n-1}}$$

$$<\frac{b(1-k)+\sum_{n=2}^{\infty}[b(1-k)(\delta+2)+2(n-1)]\frac{\Gamma(n+\delta+1)}{(n-1)!\Gamma(3+\delta)}|a_n|}{b(1-2\alpha+k)-\sum_{n=2}^{\infty}|b(1-2\alpha+k)(\delta+2)+2(n-1)|\frac{\Gamma(n+\delta+1)}{(n-1)!\Gamma(3+\delta)}|a_n|}.$$

This expression is bounded above by 1 if

$$b(1-k)+\sum_{n=2}^{\infty}[b(1-k)(\delta+2)+2(n-1)]\frac{\Gamma(n+\delta+1)}{(n-1)!\Gamma(3+\delta)}|a_n|$$
$$\leq b(2\alpha-1-k)-\sum_{n=2}^{\infty}|b(1-2\alpha+k)(\delta+2)+2(n-1)|\frac{\Gamma(n+\delta+1)}{(n-1)!\Gamma(3+\delta)}|a_n|,$$

which is equivalent to our condition as stated in (2.1) of the theorem. □

If we take $k=1$ and $1<\alpha\leq\frac{3}{2}$ in Theorem 2.1, we have

Theorem 2.2. *If $f(z)\in N$ satisfies*

$$\sum_{n=2}^{\infty}[b(1-\alpha)(\delta+2)+2(n-1)]\frac{\Gamma(n+\delta+1)}{(n-1)!\Gamma(3+\delta)}\leq b(\alpha-1) \tag{2.2}$$

for $1<\alpha\leq\frac{3}{2}$, then $f(z)\in M^{\delta}(b,\alpha)$.

The function $f(z)$ given by

$$f(z)=z+\sum_{n=2}^{\infty}\frac{1}{n(n+1)}$$
$$\times\frac{4(\alpha-1)\frac{\Gamma(n+1+\delta)}{(n-1)!\Gamma(2+\delta)}z^n}{[b(1-k)(\delta+2)+2(n-1)]+|b(1-2\alpha+k)(\delta+2)+2(n-1)|} \tag{2.3}$$

belongs to the class $M^{\delta}(b,\alpha)$.

Let A_n denote the class of functions $f(z)$ normalized in the form

$$f(z) = z + \sum_{k=n}^{\infty} a_k z^k \quad (n \in N \backslash \{1\}); \quad N = \{1,2,3\ldots\}, \tag{2.4}$$

which are analytic in the open unit disc $U = \{z : |z| < 1\}$.

We denote by $M_n^{\delta}(b,\alpha)$ the subclass of A_n consisting of functions $f(z)$ which satisfy the inequality

$$\Re\left\{1 - \frac{2}{b} + \frac{2}{b}\frac{D^{\delta+2}f(z)}{D^{\delta+1}f(z)}\right\} < 1 \qquad (\alpha > 1, z \in U). \tag{2.5}$$

It is easy to derive the following theorem, which provides the sufficient condition for functions $f \in A_n$ to be in $M_n^{\delta}(b,\alpha)$.

Theorem 2.3. *If $f \in A_n$ given by (2.4) satisfies the inequality*

$$\sum_{k=n}^{\infty} [b(1-n)(\delta+2) + 2(k-1)] \frac{\Gamma(k+\delta+1)}{(k-1)!\Gamma(3+\delta)} |a_k| \leq b(\alpha-1), \tag{2.6}$$

$\left(1 < \alpha < \frac{n+1}{2}\right)$ *then $f \in M_n^{\delta}(b,\alpha)$.*

As direct corollaries to the above theorems, we record:

Corollary 2.4. *For $\delta = -1, b = 2$ we obtain Theorem 2.1 in [1]. If $f(z) \in N$ satisfies*

$$\sum_{n=2}^{\infty} [(n-k) + |n+k-2\alpha|]|a_n| \leq 2(\alpha-1) \tag{2.7}$$

for $0 \leq k \leq 1$ and $\alpha > 1$, then $f(z) \in M(\alpha)$.

Corollary 2.5. *For $\delta = 0, b = 1$ we have Theorem 2.3 in [1]. If $f(z) \in N$ satisfies*

$$\sum_{n=2}^{\infty} n[(n-k+1) + n + k - 2\alpha]|a_n| \leq 2(\alpha-1) \tag{2.8}$$

for $0 \leq k \leq 1$ and $\alpha > 1$, then $f(z) \in N(\alpha)$.

Corollary 2.6. *For $\delta = -1, b = 2$ we get Corollary 2.2 in [1] and for $\delta = 0, b = 1$, we obtain Corollary 2.4 in [1].*

Corollary 2.7. *For $\delta = -1, b = 2$ in Theorem 2.3 we get Lemma 1.1 in [2] and for $\delta = 0, b = 1$ we obtain Lemma 1.2 in [2].*

Our results for the coefficient estimates of functions $f(z) \in M^{\delta}(b,\alpha)$, is contained in the following:

Theorem 2.8. *If $f(z) \in M^{\delta}(b,\alpha)$, then*

$$|a_n| \leq \frac{b}{2(\delta+2)} \cdot \frac{\prod_{j=2}^{n}(j+2\alpha-4)}{\Gamma(n+\delta+1)} \qquad (n \geq 2). \tag{2.9}$$

Proof. Let us define the function $p(z)$ by

$$p(z) = \frac{\alpha - \left(1 - \frac{2}{b} + \frac{2}{b}\frac{D^{\delta+2}f(z)}{D^{\delta+1}f(z)}\right)}{\alpha - 1}$$

for $f(z) \in M^{\delta}(b,\alpha)$. Then $p(z)$ is analytic $p(0) = 1$ and $\Re\{p(z)\} > 0$ $(z \in U)$.

If $p(z) = 1 + \sum_{n=1}^{\infty} p_n z^n$ then, $|p_n| \leq 2$ $(n \geq 1)$.

Since

$$(\alpha-1)bp(z)D^{\delta+1}f(z) = [(\alpha-1)+b+2]D^{\delta+1}f(z) - 2D^{\delta+2}f(z),$$

we get

$$\begin{aligned} &2(1-n)a_n \frac{\Gamma(n+\delta+1)}{(n-1)!\Gamma(3+\delta)} \\ &= (\alpha-1)b\left[p_{n-1} + p_{n-2}(\delta+2)a_2 + p_{n-3}\frac{(\delta+2)(\delta+3)a_3}{2!} + \cdots + \frac{p_1\Gamma(n+\delta)}{(n-2)!\Gamma(2+\delta)}a_{n-1}\right]. \end{aligned}$$

If $n = 2$, then $-2a_2 = (\alpha-1)bp_1$ implies that $|a_2| \leq b(\alpha-1)$. Thus the coefficient estimate (2.9) holds true for $n = 2$. Suppose that

$$|a_k| \leq \frac{b}{2(\delta+2)} \cdot \frac{\prod_{j=2}^{k}(j+2\alpha-4)}{(k-1)!}$$

is true for all $k = 2,3,4,\ldots,n$. We have

$$\begin{aligned} -2na_{n+1}\frac{\Gamma(n+\delta+2)}{n!\Gamma(3+\delta)} = (\alpha-1)b\Bigg[&p_n + p_{n-1}(\delta+2)a_2 + p_{n-2}\frac{(\delta+2)(\delta+3)}{2!}a_3 \\ &+ \cdots + p_1\frac{\Gamma(n+1+\delta)}{(n-1)!\Gamma(2+\delta)}a_n\Bigg], \end{aligned}$$

$$\begin{aligned} &2n|a_{n+1}| \\ &\leq 2(\alpha-1)b\left[1 + (\delta+2)|a_2| + \frac{(\delta+2)(\delta+3)}{2!}|a_3| + \cdots + \frac{\Gamma(n+1+\delta)}{(n-1)!\Gamma(2+\delta)}|a_n|\right] \\ &\leq \frac{b}{2(\delta+2)\Gamma(n+2+\delta)}\left[1 + (2\alpha-2) + \frac{(2\alpha-2)(2\alpha-1)}{2!} + \cdots + \frac{\prod_{j=2}^{n}(j+2\alpha-4)}{(n-1)!}\right] \\ &\leq \frac{b}{2(\delta+2)\Gamma(n+\delta+2)}\left[\frac{(2\alpha-1)2\alpha(2\alpha+1)\cdots(2\alpha+n-4)}{(n-2)!}\right. \\ &\qquad \left. + \frac{(2\alpha-2)(2\alpha-1)2\alpha\cdots(2\alpha+n-4)}{(n-1)!}\right] \\ &= \frac{b}{2(\delta+2)\Gamma(n+\delta+2)}\prod_{j=2}^{n+1}(j+2\alpha-4). \end{aligned}$$

Hence, the coefficient estimate (2.9) is true for $k = n+1$. By mathematical induction for the coefficient estimate (2.9) we get the result. □

We obtain as corollaries of Theorem 2.8 the following results.

Corollary 2.9. *For* $\delta = -1, b = 2$ *we get Theorem* 2.6 *in [1]. If* $f(z) \in M(\alpha)$, *then*

$$|a_n| \leq \frac{\prod_{j=2}^{n}(j+2\alpha-4)}{(n-1)!} \qquad (n \geq 2).$$

Corollary 2.10. *For* $\delta = 0, b = 1$ *we obtain Theorem* 2.7 *in [1]. If* $f(z) \in N(\alpha)$, *then*

$$|a_n| \leq \frac{\prod_{j=2}^{n}(j+2\alpha-4)}{n!} \qquad (n \geq 2).$$

Bibliography

[1] S. Owa and J. Nishiwaki, Coefficient estimates for certain classes of analytic functions, *J. Ineq. Pure Appl. Math.*, **3**(5)(2002), Article 72.

[2] S. Owa and H.M. Srivastava, Some generalized convolution properties associated with certain subclasses of analytic functions, *J. Ineq. Pure Appl. Math.*, **3**(3) (2002), Article 42.

[3] H. Silverman, Univalent functions with negative coefficients, *Proc. Amer Math. Soc.*, **51** (1975), 109–116.

[4] B.A. Uralegaddi, M.D. Ganigi and S.M. Sarangi, Univalent functions with positive coefficients, *Tamkang J. Math.*, **25** (1994), 225–230.

In: Advances in Inequalities for Series
Editors: S. S. Dragomir and A. Sofo, pp. 151–157
ISBN 978-1-60021-920-7

Chapter 9

Normality of Meromorphic Functions and Shared Sets

Chunlin Lei, Mingliang Fang and Degui Yang
Department of Applied Mathematics
South China Agricultural University
Guangzhou, 510642, P. R . China
E-mail addresses: leichunlin0113@126.com; hnmlfang@hotmail.com;
dyang@scau.edu.cn

Abstract

The main aim of this paper is to prove the following. Let $\mathcal{F}$ be a family of meromorphic functions in a domain D, let $n(\geq 2), k$ be two positive integers, and let $S = \{a_1, a_2, \ldots, a_n\}$, where $a_1, a_2, \ldots, a_n$ are distinct finite complex numbers. If for each $f \in \mathcal{F}$, all zeros of f have multiplicity at least $k+1$, and f and $f^{(k)}$ share the set S in D, then $\mathcal{F}$ is normal in D.

1991 Mathematics Subject Classification: 30D45

Key words and phrases: Meromorphic function, Sharing value, Unicity.

1. Introduction and Main Results

Let D be a domain in C and $\mathcal{F}$ a family of meromorphic functions defined in D. $\mathcal{F}$ is said to be normal in D, in the sense of Montel, if each sequence $\{f_n\} \subset \mathcal{F}$ has a subsequence $\{f_{n_j}\}$ which converges spherically locally uniformly in D, to a meromorphic function or ∞ (see [4, 8, 11]).

Let f and g be meromorphic functions in D, and let a be a complex number. If $f(z) = a$ if and only if $g(z) = a$, then we say that f and g share a in D.

Mues and Steinmetz [5] proved:

Theorem A. *Let f be a non-constant meromorphic function, and let a_1, a_2, and a_3 be three distinct finite complex numbers. If f and f' share a_1, a_2, and a_3, then $f \equiv f'$.*

Schwick [9] discovered a connection between normality criteria and shared values. He proved:

Theorem B. *Let $\mathcal{F}$ be a family of meromorphic functions in a domain D, and let a_1, a_2, and a_3 be three distinct finite complex numbers. If, for each $f \in \mathcal{F}$, f and f' share a_1, a_2, and a_3 in D, then $\mathcal{F}$ is normal in D.*

Pang and Zalcman [7] improved Theorem B as follows.

Theorem C. *Let $\mathcal{F}$ be a family of meromorphic functions in a domain D, and let a, b be two distinct complex numbers. If, for each $f \in \mathcal{F}$, f and f' share a and b in D, then $\mathcal{F}$ is normal in D.*

Recently, Fang and Zalcman [3] proved:

Theorem D. *Let $\mathcal{F}$ be a family of meromorphic functions in a domain D, let a be nonzero finite complex number, and let k be a positive integer. If, for every $f \in \mathcal{F}$, all zeros of f have multiplicity at least $k+1$, and f and $f^{(k)}$ share a in D, then $\mathcal{F}$ is normal in D.*

Let S be a set of complex numbers. If $f(z) \in S$ if and only if $g(z) \in S$ in a domain D, then we say that f and g share the set S in D.

It is natural to ask whether Theorem D remains valid if "f and $f^{(k)}$ share a value a" is replaced by "f and $f^{(k)}$ share a set S"?

In this note, we give an affirmative answer to this problem and prove:

Theorem 1.1. *Let $\mathcal{F}$ be a family of meromorphic functions in a domain D, let $n\,(\geq 2)$, k be two positive integers, and let $S = \{a_1, a_2, \ldots, a_n\}$, where $a_1, a_2, \ldots, a_n$ are distinct finite complex numbers. If, for every $f \in \mathcal{F}$, all zeros of f have multiplicity at least $k+1$, and f and $f^{(k)}$ share the set S in D, then $\mathcal{F}$ is normal in D.*

Example 1. *Let $S = \{1, -1\}$. Set*

$$\mathcal{F} = \{f_n(z) : n = 2, 3, 4, \ldots\},$$

where

$$f_n(z) = \frac{n+1}{2n} e^{nz} + \frac{n-1}{2n} e^{-nz}, \quad D = \{z : |z| < 1\}.$$

Then, for any $f_n \in \mathcal{F}$, we have

$$n^2[f_n^2(z) - 1] = [f_n'(z)]^2 - 1.$$

Thus f and f' share the set S in D, but $\mathcal{F}$ is not normal in D. This shows that all zeros of f having multiplicity at least $k+1$ are best possible (at least for $k = 1$).

Theorem 1.2. *Let $\mathcal{F}$ be a family of meromorphic functions in a domain D, let n, m, k be three positive integers, and let $S_1 = \{a_1, a_2, \ldots, a_n\}$, $S_2 = \{b_1, b_2, \ldots, b_m\}$ where $a_1, a_2, \ldots, a_n, b_1, b_2, \ldots, b_m$ are distinct finite complex numbers. If, for each $f \in \mathcal{F}$, all zeros of f have multiplicity at least k, and f and $f^{(k)}$ share the sets S_1 and S_2 in D, then $\mathcal{F}$ is normal in D.*

Example 2. *Fix k and let* $\{\omega_1, \omega_2, \ldots, \omega_k\}$ *be the k-th roots of unity* $(\omega^k = 1)$*. Any function of the form*

$$F(z) = \sum_{j=1}^{k} c_j e^{\omega_j z}$$

clearly satisfies $F^{(k)} \equiv F$*. The* $k \times k$ *Vandermonde determinant defined by* ω_j, $1 \le j \le k$, *does not vanish. Hence, solving k linear equations in k unknowns, we may choose the* c_j *so that the first* $k-1$ *Taylor coefficients of F vanish at the origin, i.e., so that F has a zero of order* $k-1$ *at 0. Let* $D = \{z : |z| < 1\}$*, and set* $f_n(z) = nF(z), n = 1, 2, 3, \ldots$ *. Let* $\mathcal{F} = \{f_n\}$*; then* $\mathcal{F}$ *is a family of holomorphic functions on D. Obviously, for each* $f \in \mathcal{F}$, $f^{(k)} \equiv f$*, so f and* $f^{(k)}$ *share every set in D. But* $\mathcal{F}$ *is not normal in D. This shows that the requirement of multiplicity k in Theorem 1.2 cannot be dropped in general.*

2. Some Lemmas

For the proof of our theorems, we need the following lemmas.

Lemma 2.1 ([6, Lemma 2], [12]). *Let* $\mathcal{F}$ *be a family of functions meromorphic in the unit disc, all of whose zeros have multiplicity at least k, and suppose that there exists* $A \ge 1$ *such that* $|f^{(k)}(z)| \le A$ *whenever* $f(z) = 0$*. Then, if* $\mathcal{F}$ *is not normal, there exist, for each* $0 \le \alpha \le k$,

a) a number $0 < r < 1$;

b) points z_n, $|z_n| < r$;

c) functions $f_n \in \mathcal{F}$*; and*

d) positive numbers $\rho_n \to 0^+$

such that $\rho_n^{-\alpha} f_n(z_n + \rho_n \zeta) = g_n(\zeta) \to g(\zeta)$ *locally uniformly with respect to the spherical metric, where g is a nonconstant meromorphic function on C, all of whose zeros have multiplicity at least k, such that* $g^{\#}(\zeta) \le g^{\#}(0) = kA + 1$*. In particular, g has order at most 2.*

Here, as usual, $g^{\#}(\zeta) = |g'(\zeta)|/(1 + |g(\zeta)|^2)$ is the spherical derivative.

Lemma 2.2 ([1], [10, Lemma 6]). *Let f be a transcendental meromorphic function of finite order, let a be a nonzero complex number, and let k be a positive integer. If all zeros of f have multiplicity at least* $k+1$*, then* $f^{(k)} - a$ *has infinitely many zeros.*

Lemma 2.3 ([10, Lemma 8]). *Let* $f(z) = a_n z^n + a_{n-1} z^{n-1} + \cdots + a_0 + q(z)/p(z)$*, where* a_0, $a_1, \ldots, a_n$ *are constants with* $a_n \ne 0$*, and q and p are two co-prime polynomials neither of which vanishes identically, with* $\deg q < \deg p$*, and let k be a positive integer. If the zeros of f have multiplicity at least* $k+1$*, and* $f^{(k)} \ne 1$*, then*

$$f(z) = \frac{z^k}{k!} + \cdots + a_0 + \frac{1}{az+b}.$$

Here $a (\ne 0)$, b, $a_0, \ldots$ *are constants.*

3. Proofs of Theorems 1.1 – 1.2

Proof (Proof of Theorem 1.1). We may assume that $S = \{a,b\}$, where a and b are two distinct constants and $D = \triangle$, the unit disc. Now we consider two cases:

Case 1. $ab \neq 0$. Suppose that $\mathcal{F}$ is not normal in $\triangle$. Then by Lemma 2.1, we can find $f_n \in \mathcal{F}$, $z_n \in \triangle$ and $\rho_n \to 0^+$ such that $g_n(\zeta) = \rho_n^{-k} f_n(z_n + \rho_n \zeta)$ converges locally uniformly with respect to the spherical metric to a nonconstant meromorphic function g on C, all of whose zeros have multiplicity at least $k+1$.

We claim that

(i) $g^{(k)} \neq a$;

(ii) $g^{(k)} \neq b$.

Suppose that $g^{(k)}(\zeta_0) = a$. We claim that $g^{(k)} \not\equiv a$, for otherwise g would be a polynomial of degree at most k, and so could not have zeros of multiplicity at least $k+1$. Since $g_n^{(k)}(\zeta) \to g^{(k)}(\zeta)$ on a neighborhood of ζ_0, there exist ζ_n, $\zeta_n \to \zeta_0$, such that (for n sufficiently large)

$$a = g^{(k)}(\zeta_0) = g_n^{(k)}(\zeta_n) = f_n^{(k)}(z_n + \rho_n \zeta_n).$$

Thus $f_n^{(k)}(z_n + \rho_n \zeta_n) = a$, so that $f_n(z_n + \rho_n \zeta_n) = a$ or b. It follows that $g_n(\zeta_n) = \rho_n^{-k} f_n(z_n + \rho_n \zeta_n)$, and so $g(\zeta_0) = \lim_{n \to \infty} g_n(\zeta_n) = \infty$, which contradicts $g^{(k)}(\zeta_0) = a$.

In a similar fashion, we can prove that $g^{(k)}(\zeta_0) \neq b$. This completes the proof of (i) and (ii).

By Lemma 2.2 and (i), we know that g is not a transcendental meromorphic function. Since g has zeros of multiplicity at least $k+1$ and $g^{(k)} \neq a$, it follows that g is not a polynomial. Hence by (i) and Lemma 2.3, we obtain that $g(\zeta) = \frac{a\zeta^k}{k!} + \cdots + a_0 + \frac{1}{A\zeta + B}$, where B, $A(\neq 0)$, $a_0, \ldots$ are constants. Thus

$$g^{(k)} = a + \frac{(-1)^k k! A^k}{(A\zeta + B)^{k+1}}.$$

It follows that $g^{(k)}(\zeta) = b$ has solutions, which contradicts (ii). Hence $\mathcal{F}$ is normal in D.

Case 2. $ab = 0$, we may assume that $b = 0$. Suppose that $\mathcal{F}$ is not normal in $\triangle$. Then by Lemma 2.1, we can find $f_n \in \mathcal{F}$, $z_n \in \triangle$ and $\rho_n \to 0^+$ such that $g_n(\zeta) = \rho_n^{-k} f_n(z_n + \rho_n \zeta)$ converges locally uniformly with respect to the spherical metric to a nonconstant meromorphic function g on C, all of whose zeros have multiplicity at least $k+1$.

We claim that:

(iii) $g^{(k)} \neq a$;

(iv) $g^{(k)} = 0 \Leftrightarrow g = 0$.

Suppose that $g^{(k)}(\zeta_0) = a$. We claim that $g^{(k)} \not\equiv a$, for otherwise g would be a polynomial of degree at most k, and so could not have zeros of multiplicity at least $k+1$. Since $g_n^{(k)}(\zeta) \to g^{(k)}(\zeta)$ on a neighborhood of ζ_0, there exist ζ_n, $\zeta_n \to \zeta_0$, such that (for n sufficiently large)

$$a = g^{(k)}(\zeta_0) = g_n^{(k)}(\zeta_n) = f_n^{(k)}(z_n + \rho_n \zeta_n).$$

Thus $f_n^{(k)}(z_n+\rho_n\zeta_n)=a$, so that $f_n(z_n+\rho_n\zeta_n)=a$. It follows that $g_n(\zeta_n)=\rho_n^{-k}f_n(z_n+\rho_n\zeta_n)=\rho_n^{-k}a$, and so $g(\zeta_0)=\lim_{n\to\infty}g_n(\zeta_n)=\infty$, which contradicts $g^{(k)}(\zeta_0)=a$. This completes the proof of (iii).

Next, we prove (iv). Suppose that $g^{(k)}(\zeta_0)=0$. Clearly $g^{(k)}\not\equiv 0$, for otherwise g would be a polynomial of degree less than k, and so could not have zeros of multiplicity at least $k+1$. Since $g_n^{(k)}(\zeta)\to g^{(k)}(\zeta)$ on a neighborhood of ζ_0, there exist ζ_n, $\zeta_n\to\zeta_0$, such that (for n sufficiently large)

$$0=g^{(k)}(\zeta_0)=g_n^{(k)}(\zeta_n)=f_n^{(k)}(z_n+\rho_n\zeta_n).$$

Thus $f_n^{(k)}(z_n+\rho_n\zeta_n)=0$, so that $f_n(z_n+\rho_n\zeta_n)=0$ or a. Then, for sufficiently large n, we have $f_n(z_n+\rho_n\zeta_n)=a$ or $f_n(z_n+\rho_n\zeta_n)=0$.

If $f_n(z_n+\rho_n\zeta_n)=a$, then $g(\zeta_0)=\lim_{n\to\infty}g_n(\zeta_n)=\infty$, which contradicts $g^{(k)}(\zeta_0)=0$.

Hence $f_n(z_n+\rho_n\zeta_n)=0$, so that $g(\zeta_0)=\lim_{n\to\infty}g_n(\zeta_n)=0$. Thus we deduce that $g^{(k)}=0\Rightarrow g=0$. Obviously, $g=0\Rightarrow g^{(k)}=0$. This proves (iv).

By Lemma 2.2 and (iii), we know that g is not a transcendental meromorphic function. Since g has zeros of multiplicity at least $k+1$ and $g^{(k)}\neq a$, it follows that g is not a polynomial. By Lemma 2.3, we have

$$g(\zeta)=\frac{a}{k!}\zeta^k+\cdots+a_0+\frac{1}{A\zeta+B},$$

where $B,A(\neq 0),a_0,\ldots$ are constants. Thus $g(\zeta)$ has only a zero. On the other hand, $g^{(k)}=a+\frac{(-1)^k k!A^k}{(A\zeta+B)^{k+1}}$. Obviously, $g^{(k)}=0$ has $k+1$ distinct solutions, which contradicts $g^{(k)}=0\Rightarrow g=0$. Hence $\mathcal{F}$ is normal in D. The proof of Theorem 1.1 is complete. □

Proof (Proof of Theorem 1.2). We may assume that $S_1=\{a,b\},S_2=\{c\}$, where a,b,c are three distinct constants and $D=\triangle$, the unit disc. Suppose that $\mathcal{F}$ is not normal in $\triangle$. Then by Lemma 2.1, we can find $f_n\in\mathcal{F}$, $z_n\in\triangle$ and $\rho_n\to 0^+$ such that $g_n(\zeta)=f_n(z_n+\rho_n\zeta)$ converges locally uniformly with respect to the spherical metric to a nonconstant meromorphic function g on C, all of whose zeros have multiplicity at least k.

We claim that:

(i) $g^{(k)}\neq 0$;

(ii) $g\neq a, g\neq b$, and $g\neq c$.

Suppose that $g^{(k)}(\zeta_0)=0$. Clearly $g^{(k)}\not\equiv 0$, for otherwise g would be a polynomial of degree less than k, and so could not have zeros of multiplicity at least k. Since $g_n^{(k)}(\zeta)-\rho_n^k a\to g^{(k)}(\zeta)$ on a neighborhood of ζ_0, there exist ζ_n, $\zeta_n\to\zeta_0$, such that (for n sufficiently large)

$$0=g^{(k)}(\zeta_0)=g_n^{(k)}(\zeta_n)-\rho_n^k a=\rho_n^k(f_n^{(k)}(z_n+\rho_n\zeta_n)-a).$$

Thus $f_n^{(k)}(z_n+\rho_n\zeta_n)=a$, so that $f_n(z_n+\rho_n\zeta_n)=a$ or b. It follows that $g_n(\zeta_n)=f_n(z_n+\rho_n\zeta_n)=a$ or b, and so $g(\zeta_0)=\lim_{n\to\infty}g_n(\zeta_n)=a$ or b. In a similar fashion, using

$g_n^{(k)}(\zeta)-\rho_n c$ instead of $g_n^{(k)}(\zeta)-\rho_n a$, we obtain that $g(\zeta_0)=c$, which contradicts $a\neq c$ and $b\neq c$. This completes the proof of (i).

Next, we prove (ii). Suppose that $g(\zeta_0)=a$. Then there exist $\zeta_n\to\zeta_0$, such that (for n sufficiently large)

$$a=g(\zeta_0)=g_n(\zeta_n)=f_n(z_n+\rho_n\zeta_n).$$

Thus $f_n(z_n+\rho_n\zeta_n)=a$, so that $f_n^{(k)}(z_n+\rho_n\zeta_n)=a$ or b. It follows that $g_n^{(k)}(\zeta_n)=\rho_n^k f_n^{(k)}(z_n+\rho_n\zeta_n)\to 0$. Therefore, $g^{(k)}(\zeta_0)=\lim_{n\to\infty} g_n^{(k)}(\zeta_n)=0$, which contradicts (i).

In a similar fashion, we obtain that $g\neq b$ and $g\neq c$.

Now by Picard's Theorem and (ii), g is a constant, a contradiction. Thus $\mathcal{F}$ is normal in D. The theorem is proved. □

Bibliography

[1] W. Bergweiler and A. Eremenko, On the singularities of the inverse to a meromorphic function of finite order, *Rev. Mat. Iber.*, **11** (1995), 355–373.

[2] M.L. Fang and L. Zalcman, Normal families and shared values of meromorphic functions, *Ann. Polon. Math.*, **80** (2003), 133–141.

[3] M.L. Fang and L. Zalcman, Normal families and shared values of meromorphic functions III , *Comput. Methods Funct. Theory*, **2** (2002), 385–395.

[4] W.K. Hayman, *Meromorphic Functions*, Clarendon Press, Oxford, 1964.

[5] E. Mues and N. Steinmetz, Meromorphe Funktionen die mit ihrer Ableitung Werte teilen, *Manuscripta Math.*, **29** (1979), 195–206.

[6] X.C. Pang and L. Zalcman, Normal families and shared values, *Bull. London Math. Soc.*, **32** (2000), 325–331.

[7] X.C. Pang and L. Zalcman, Normality and shared values, *Ark. Mat.*, **38** (2000), 171–182.

[8] J. Schiff, *Normal Families*, Springer-Verlag, 1993.

[9] W. Schwick, Sharing values and normality, *Arch. Math.*, **59** (1992), 50–54.

[10] Y.F. Wang and M.L. Fang, Picard values and normal families of meromorphic functions with zeros, *Acta Math. Sinica, New Series*, **14**(1) (1998), 17–26.

[11] L. Yang, *Value Distribution Theory*, Springer-Verlag, Berlin, 1993.

[12] L. Zalcman, Normal families: new perspectives, *Bull. Amer. Math. Soc.*, **35** (1998), 215–230.

[13] W.H. Zhang, The normality of meromorphic functions concerning sharing set, *J. Nanhua University* (Science and Technology), **18**(4) (2004), 36–38.

[14] W.H. Zhang, Normal families of meromorphic functions, J. *East China University of Science and Technology*, **30**(6) (2004), 709–711.

In: Advances in Inequalities for Series
Editors: S. S. Dragomir and A. Sofo, pp. 159–167
ISBN 978-1-60021-920-7

Chapter 10

Equivalence Conditions Regarding Numerical Sequences and Their Applications

L. Leindler
University of Szeged, Aradi vértanúk tere 1
H-6720 Szeged, Hungary
E-mail address: leindler@math.u-szeged.hu

Abstract

We prove two equivalence theorems regarding almost decreasing numerical sequences. As an application we also present estimations of new type for best approximation of functions having coefficients of bounded variation by trigonometric polynomials in $L^p_{2\pi}$-space.

1991 Mathematics Subject Classification: 42A10, 42A32, 40A05, 26D15.

Key words and phrases: Numerical sequences, Best approximation, Sine and cosine series, Equivalence conditions.

1. Introduction

In [4] we cited many known theorems pertaining to sine or cosine series with positive decreasing or quasi-decreasing coefficients. In the same paper we extended five theorems of them for such series where the coefficients belong to the sequence class of rest bounded variation, in symbol to the class of $RBVS$, that is, $\mathbf{c} := \{c_n\} \in RBVS$ if $c_n \to 0$ and

$$\sum_{n=m}^{\infty} |\Delta c_n| \le K(\mathbf{c})c_m, \quad \Delta c_n := c_n - c_{n+1}, \quad K(\mathbf{c}) > 0, \quad m = 0, 1, \dots .$$

In [5] we defined a wider class of sequences. This class is denoted by $\gamma RBVS$, where $\gamma := \{\gamma_n\}$ is a positive sequence, and a null-sequence of *real numbers* $\mathbf{c} \in \gamma RBVS$ if for all m

$$\sum_{n=m}^{\infty} |\Delta c_n| \le K(\mathbf{c})\gamma_m$$

holds.

In [6] we determined a further class. Its definition is the following: A sequence $\mathbf{c}$ of real numbers belongs to the class BVS_0 if $c_n \to 0$ and $\sum_{n=1}^{\infty} |\Delta c_n| < \infty$.

We note that if the sequence $\gamma := \{\gamma_n\}$ has the terms $\gamma_n := \sum_{k=n}^{\infty} |\Delta c_k|$ $(< \infty)$, then the classes $\gamma RBVS$ and BVS_0 are identical. Furthermore, it is easy to see that if $\mathbf{c} \in RBVS$ then it is also *almost monotonic*, that is, for all $n \geq m$

$$(0 \leq) c_n \leq K(\mathbf{c}) c_m, \tag{1.1}$$

but (1.1) does not hold if $\mathbf{c} \in \gamma RBVS$ $(\gamma \neq c)$ or $\mathbf{c} \in BVS_0$. In general, the class BVS_0 is the widest one, its sequences may have zero and negative terms as well.

The aim of this paper is to verify two equivalence results initiated by theorems of A.A. Konyushkov [3, Theorem 8 and 9]. Utilizing one of these results we present estimations for the best approximation of functions having coefficients belonging to the class BVS_0.

In order to formulate our results we need some definitions, notations and two theorems.

Let $f(x) \in L^p_{2\pi}$, $1 < p < \infty$, and let $E_n(p, f)$ denote the best approximation of f by trigonometric polynomials of order at most n in $L^p_{2\pi}$-space.

We shall use the notation $L \ll R$ at inequalities if there exists a positive constant K such that $L \leq KR$ holds.

If $\rho := \{\rho_n\}$ and $\eta := \{\eta_n\}$ are positive sequences such that

$$\eta_n \ll \rho_n \ll \eta_n$$

hold for all n, then we shall use the notation: $\rho_n \sim \eta_n$.

It will be said that a positive function $\varphi(\delta)$ tending to 0 as $\delta \to 0$ satisfies the "γ-*condition*" if there exists a γ, $0 < \gamma < 1$, such that $\varphi(\delta) \ll \varphi(\gamma\delta)$ holds.

If a function $\varphi(\delta)$ has the property

$$\liminf_{\delta \to 0} \frac{\varphi(C\delta)}{\varphi(\delta)} > 1$$

for some $C > 1$, then we say that it satisfies the "*L-condition*". For further details on these conditions, consult [1] and [2].

An almost decreasing sequence $\{\gamma_n\}$ of positive terms is *quasi geometrically decreasing* if there exists a natural number μ such that

$$\gamma_{n+\mu} \leq \frac{1}{2}\gamma_n, \qquad n = 1, 2, \ldots .$$

A sequence $\{\gamma_n\}$ of positive terms is *quasi* β-*power-monotone decreasing* if there exists a constant $K = K(\beta, \{\gamma_n\}) \geq 1$ such that

$$n^\beta \gamma_n \leq K m^\beta \gamma_m$$

holds for any $n \geq m$.

In order to present an application of our equivalence theorem we have to recall the following two theorems proved in [6].

Theorem A. *Let* $1 < p < \infty$, $\mathbf{b} := \{b_n\} \in BVS_0$, $\gamma := \{\gamma_n := \sum_{k=n}^{\infty} |\Delta b_k|\}$ *and*

$$\sum_{n=1}^{\infty} \gamma_n^p n^{p-2} < \infty. \tag{1.2}$$

If $g(x)$ *denotes the sum of the series*

$$\sum_{n=1}^{\infty} b_n \sin nx, \tag{1.3}$$

then $g \in L_{2\pi}^p$ *and*

$$E_n(p,g) \ll \left\{ \gamma_{n+1}(n+1)^{\frac{1}{p'}} + \left(\sum_{k=n+1}^{\infty} \gamma_k^p k^{p-2} \right)^{\frac{1}{p}} \right\}, \tag{1.4}$$

where $p' := p/(p-1)$.

If $\gamma_n = b_n$, then Theorem A reduces to Theorem 1 in [4], which is a generalization of a theorem of Konyushkov [3, Theorem 4].

Theorem B. *Using the notations of Theorem A, and if*

$$\sum_{n=1}^{\infty} n^{-1/p} \gamma_n < \infty, \tag{1.5}$$

then $g \in L_{2\pi}^p$ *and*

$$E_n(p,g) \ll \left\{ \gamma_{n+1}(n+1)^{\frac{1}{p'}} + \sum_{k=n+1}^{\infty} k^{-1/p} \gamma_k \right\}. \tag{1.6}$$

This theorem with $\mathbf{b} \in RBVS$ was proved in [4], and for quasi-monotone sequences in [3].

We mention that analogous results for cosine series can be proved, too.

2. Theorems

Theorem 2.1. *Let* $\gamma(\delta)$ *be a positive function defined on* $[0,1]$, *nondecreasing and satisfying the conditions* (γ), (L) *and* $\varphi(\delta) \to 0$, *as* $\delta \to 0$. *If* $\rho := \{\rho_n\}$ *is an almost decreasing numerical sequence,* $1 < q < p < \infty$, $\varphi_n := \varphi(n^{-1})$, *then the following six relations are pairwise equivalent:*

$$n^{1-1/p} \rho_n \ll \varphi_n, \tag{2.1}$$

$$\sum_{k=n}^{\infty} k^{-1/p} \rho_k \ll \varphi_n, \tag{2.2}$$

$$\sum_{k=n}^{\infty} \rho_k^p k^{p-2} \ll \varphi_n^p, \tag{2.3}$$

$$\sum_{k=n}^{\infty} \rho_k^{p'} \ll \varphi_n^{p'}, \qquad \left(p' = \frac{p}{p-1}\right), \tag{2.4}$$

$$\sum_{k=n}^{\infty} k^{-1/q} \rho_k \ll n^{1/p-1/q} \varphi_n, \tag{2.5}$$

$$\sum_{k=n}^{\infty} \rho_k^q k^{q-2} \ll n^{q/p-1} \varphi_n^q. \tag{2.6}$$

Theorem 2.2. *If p, q and φ_n have the same meaning as in Theorem 2.1,* **b** *and γ as in Theorem A, furthermore if $n^{1-1/p}\gamma_n \ll \varphi_n$, then $g \in L_{2\pi}^p$,*

$$E_n(p,g) \ll \varphi_n \tag{2.7}$$

and

$$E_n(q,g) \ll n^{1/p-1/q} \varphi_n \tag{2.8}$$

hold.

Theorem 2.3. *If $p, p', q, \rho, \varphi(\delta)$ and φ_n have the same meaning as in Theorem 2.1, then the following six relations are pairwise equivalent:*

$$n^{1-1/p} \rho_n \sim \varphi_n, \tag{2.9}$$

$$\sum_{k=n}^{\infty} k^{-1/p} \rho_k \sim \varphi_n, \tag{2.10}$$

$$\sum_{k=n}^{\infty} \rho_k^p k^{p-2} \sim \varphi_n^p, \tag{2.11}$$

$$\sum_{k=n}^{\infty} \rho_k^{p'} \sim \varphi_n^{p'}, \tag{2.12}$$

$$\sum_{k=n}^{\infty} k^{-1/q} \rho_k \sim n^{1/p-1/q} \varphi_n, \tag{2.13}$$

$$\sum_{k=n}^{\infty} \rho_k^q k^{q-2} \sim n^{q/p-1} \varphi_n^p. \tag{2.14}$$

3. Lemmas

Lemma 3.1 (see [1]). *If $\varphi(\delta)$ given in Theorem 2.1 satisfies condition (L), then for any $p \geq 1$ and all $n \in \mathbb{N}$*

$$\sum_{k=n}^{\infty} k^{-1}\varphi_k^p \ll \varphi_n^p. \tag{3.1}$$

Lemma 3.2. *Let $\varphi(\delta)$ have the same meaning and property as in Lemma 3.1. Then there exists a natural number μ such that for all n*

$$\varphi_{2^{n+\mu}} < \frac{1}{2}\varphi_{2^n} \tag{3.2}$$

holds.

Proof. Condition (L) implies that there exist N and μ with the properties

$$\varphi\left(\frac{2^\mu}{2^k}\right) > \varphi\left(\frac{C^N}{2^k}\right) > 2\varphi\left(\frac{1}{2^k}\right),$$

whence (3.2) holds with $k = \mu + n$. □

Lemma 3.3. *If a positive sequence $\{\gamma_n\}$ is almost monotonic and its partial sequence $\{\gamma_{2^n}\}$ is quasi geometrically decreasing, then the whole sequence $\{\gamma_n\}$ is quasi β-power-monotone decreasing with a certain positive β.*

This is a special case of Theorem 2.3 proved in [7].

4. Proofs of the Theorems

First we show that (2.1) implies all of the others. Let us start with the proof of the implication (2.1)⇒(2.2). Using (2.1) and (3.1) we get that

$$\sum_{k=n}^{\infty} k^{-1/p}\rho_k \ll \sum_{k=n}^{\infty} k^{-1/p}k^{1/p-1}\varphi_k \ll \varphi_k,$$

as desired.

The proofs of (2.1)⇒(2.ν), $\nu = 3,4,5$ and 6, all use (3.1), thus e.g. the proof of (2.1)⇒(2.3) runs as follows:

$$\sum_{k=n}^{\infty} \rho_k^p k^{p-2} \ll \sum_{k=n}^{\infty} k^{1-p}\varphi_k^p k^{p-2} \ll \varphi_n^p.$$

In the proofs of (2.1)⇒(2.5) and (2.1)⇒(2.6) we have to use the inequality $q < p$, too. Then

$$\sum_{k=n}^{\infty} k^{-1/q}\rho_k \ll \sum_{k=n}^{\infty} k^{-1/q}k^{1/p-1}\varphi_k \ll n^{1/p-1/q}\sum_{k=n}^{\infty} k^{-1}\varphi_k \ll n^{1/p-1/q}\varphi_n,$$

and

$$\sum_{k=n}^{\infty} \rho_k^q k^{q-2} \ll \sum_{k=n}^{\infty} k^{q/p-q} \varphi_k^q k^{q-2} \ll n^{q/p-1} \sum_{k=n}^{\infty} k^{-1} \varphi_k^q \ll n^{q/p-1} \varphi_n^q.$$

Before proving the opposite implication (2.2)⇒(2.1) we show that the property (γ) of $\varphi(\delta)$ implies that

$$\varphi_n \ll \varphi_{2n} \tag{4.1}$$

holds. Namely if N is a natural number such that $\gamma^N < 1/2$, then

$$\varphi\left(\frac{1}{n}\right) \ll \varphi\left(\gamma^N \frac{1}{n}\right) \ll \varphi\left(\frac{1}{2n}\right),$$

herewith (4.1) is verified.

Applying (4.1) the proof of (2.2)⇒(2.1) is an elementary calculation. Namely, by (4.1), we obtain that

$$\varphi_n \gg \varphi\left(\frac{2}{n}\right) \gg \sum_{k \geq n/2}^{n} k^{-1/p} \rho_k \gg \rho_n n^{1-1/p}.$$

Hereafter the implication (2.2)⇒(2.3) can be proved by the chain (2.2)⇒(2.1)⇒(2.3).

The proof of (2.3)⇒(2.4) runs similarly:

$$\varphi_n^p \gg \sum_{k \geq n/2}^{n} \rho_k^p k^{p-2} \gg \rho_n^p n^{p-1},$$

thus the chain (2.3)⇒(2.1)⇒(2.4) conveys the proof.

The proofs of (2.4)⇒(2.5) and (2.5)⇒(2.6) follow the same line, therefore we omit them.

Finally we verify (2.6)⇒(2.1). This comes from the inequalities:

$$n^{q/p-1} \varphi_n^q \gg \sum_{k \geq n/2}^{n} \rho_k^q k^{q-2} \gg \rho_n^q n^{q-1}.$$

Herewith the implication chain

$$(2.1) \Rightarrow (2.2) \Rightarrow (2.3) \Rightarrow (2.4) \Rightarrow (2.5) \Rightarrow (2.6) \Rightarrow (2.1)$$

is "closed". Consequently the pairwise equivalence of the statements (2.1)-(2.6) is verified.

The proof of Theorem 2.1 is complete.

Proof (Proof of Theorem 2.2). The inequality (2.7) comes easily using the estimate (1.4) and the inequalities (2.1) and (2.3) with γ_n instead of ρ_n.

The statement (2.8) can be proved similarly, using the estimate (1.6) and the inequalities (2.1) and (2.2) with γ_n instead of ρ_n.

The assertion $g \in L_{2\pi}^p$ is a consequence of Theorem A, namely (1.2) holds by $n^{1-1/p}\gamma_n \ll \varphi_n$ and (2.3).

The proof is complete. □

Proof (Proof of Theorem 2.3). It is easy to see that if ε is an arbitrary positive number, then (3.2), repeating it sometimes, implies the inequality

$$\varphi_{\mu n} < \varepsilon \varphi_n \tag{4.2}$$

for all n with a natural number $\mu = \mu(\varepsilon)$ being large enough. This inequality will play a crucial role in the second part of the proofs.

It is also clear that, by Theorem 2.1, it is sufficient to verify the statements only when the signs $\gg$ are considered. The proof of method will be analogous to that of Theorem 2.1, namely first we show that (2.9) implies all of the others. The proof of the pairwise equivalence takes place through (2.9).

Let us start with the proof of the implication (2.9)⇒(2.10). By (2.9)

$$\varphi_n \ll n^{1-1/p}\rho_n \ll \varphi_n. \tag{4.3}$$

In view of (2.2) it suffices to show that

$$\varphi_n \ll \sum_{k=n}^{\infty} k^{-1/p}\rho_k =: \sigma_n \tag{4.4}$$

holds. Using the inequalities (4.3) and (4.1) in due course, we get that

$$\sigma_n \geq \sum_{k=n}^{2n} k^{-1/p}\rho_k \gg \sum_{k=n}^{2n} k^{-1}\varphi_k \geq \varphi_{2n} \gg \varphi_n; \tag{4.5}$$

thus (4.4), (4.5) and (2.2) imply (2.10).

In the proofs of the implication (2.9)⇒(2.ν), $\nu = 11, 12, 13, 14$, an analogous reasoning gives the wanted assertion, therefore we omit the details.

Next we verify the implication (2.10)⇒(2.9). Now we know that

$$\varphi_n \ll \sum_{k=n}^{\infty} k^{-1/p}\rho_k \ll \varphi_n. \tag{4.6}$$

The right-hand side of (4.6) and (4.2) yield that

$$\sum_{k=\mu n}^{\infty} k^{-1/p}\rho_k < \varepsilon\varphi_n \tag{4.7}$$

holds for any $\varepsilon > 0$ and $\mu = \mu(\varepsilon)$ appearing in (4.2).

This, assuming that ε is small enough, and the left-hand side of (4.6) imply that

$$\varphi_n \ll \sum_{k=n}^{\mu n} k^{-1/p}\rho_k := H_{n,\mu}$$

also holds. Now $H_{n,\mu}$ can be estimated easily, thus

$$\varphi_n \ll H_{n,\mu} \ll \rho_n n^{1-1/p}$$

maintains. Having regard to (2.1) we have proved (2.10)⇒(2.9).

Utilizing this idea, cutting down the tail of the sums, we can prove all of the implications (2.ν)⇒(2.9), $\nu = 11, \ldots, 14$. We underline that in the proof of (2.13)⇒(2.9) at the estimation of the tail appears a factor, too. But $1/p - 1/q < 0$, thus

$$\sum_{k=\mu n}^{\infty} k^{-1/q} \rho_k \ll (\mu n)^{1/p-1/q} \varphi_{\mu n} < n^{1/p-1/q} \varphi_{\mu n} \ll n^{1/p-1/q} \varepsilon \varphi_n$$

also holds, consequently

$$n^{1/p-1/q} \varphi_n \ll \sum_{k=n}^{\mu n} \rho_n \ll \rho_n n^{1-1/q}$$

gives

$$\varphi_n \ll \rho_n n^{1-1/p},$$

as required.

The proof of (2.14)⇒(2.9) runs similarly since $q/p - 1 < 0$, too.

Collecting our results the implication chain (2.9) ⇒ (2.10) ⇒ (2.11) ⇒ (2.12) ⇒ (2.13) ⇒ (2.14) ⇒ (2.9) is closed.

Herewith the pairwise equivalence is also verified. The proof is complete. □

Bibliography

[1] N.K. Bari, On the best approximation by trigonometric polynomials of two conjugate functions, *Izv. Acad. SSSR*, **19** (1955), 285–302 (in Russian).

[2] N.K. Bari and S.B. Stečkin, Best approximation and differential properties of two conjugate functions, *Trudy Mosk. Mat. Obs.*, **5** (1956), 483–522 (in Russian).

[3] A.A. Konyushkov, Best approximation by trigonometric polynomials and Fourier coefficients, *Mat. Sb.*, **44** (1958), 53–84 (in Russian).

[4] L. Leindler, Best approximation and Fourier coefficients, *Analysis Math.*, **31** (2005), 117–129.

[5] L. Leindler, Embedding results regarding strong approximation of sine series, *Acta Sci. Math. (Szeged)*, **71** (2005), 91–103.

[6] L. Leindler, A note on the best approximation of sine and cosine series, *Analysis Math.*, **32** (2006),

[7] L. Leindler and J. Németh, On the connection between quasi power-monotone and quasi geometrical sequences with application to integrability theorems for power series, *Acta Math. Hungar.*, **68**(1-2) (1995), 7–19.

In: Advances in Inequalities
Editors: S. S. Dragomir and A. Sofo, pp. 169–179
ISBN 978-1-60021-920-7

Chapter 11

Uniqueness Theorems for Meromorphic Functions Concerning Fixed-Points

Chao Meng
Department of Mathematics, Shandong University
Jinan 250100, People's Republic of China
E-mail address: mengchaosdu@yahoo.com.cn

Abstract

In this article, we deal with the uniqueness problems on meromorphic functions concerning differential polynomials that share fixed-points. Moreover, we improve a former result.

1991 Mathematics Subject Classification: 30D35

Key words and phrases: Meromorphic function differential polynomials.

1. Introduction, Definitions and Results

In this paper, the term "meromorphic" will always mean meromorphic in the complex plane C. Let a be a complex number and $\alpha(z)$ be a meromorphic function such that $T(r,\alpha) = o(T(r,f))$, we say f and g share the value a CM, if $f-a$ and $g-a$ assume the same zeros with the same multiplicity; if $f(z)-\alpha(z)$ and $g(z)-\alpha(z)$ assume the same zeros with the same multiplicities, then we say $f(z)$ and $g(z)$ share $\alpha(z)$ CM, especially, we say that $f(z)$ and $g(z)$ have the same fixed-points when $\alpha(z) = z$. It is assumed that the reader is familiar with the notations of Nevanlinna theory, that can be found, for instance, in [1].

It is well known that if f and g share four distinct values CM, then f is a fractional transformation of g. In 1997, corresponding to one famous question of Hayman, Yang and Hua showed that similar conclusions hold for certain types of differential polynomials when they share only one value. They proved the following result.

Theorem A ([2]). *Let f and g be two nonconstant meromorphic functions, $n \geq 11$ an integer and $a \in C - \{0\}$. If $f^n f'$ and $g^n g'$ share the value a CM, then either $f = dg$ for some $(n+1)$th root of unity d or $g(z) = c_1 e^{cz}$ and $f(z) = c_2 e^{-cz}$, where c, c_1 and c_2 are constants and satisfy $(c_1 c_2)^{n+1} c^2 = -a^2$.*

In 2001, Fang and Hong obtained the following result.

Theorem B ([3]). *Let f and g be two transcendental entire functions, $n \geq 11$ an integer. If $f^n(f-1)f'$ and $g^n(g-1)g'$ share the value 1 CM, then $f \equiv g$.*

Recently, W. C. Lin and H. X. Yi extended the above theorem in view of the fixed-point. They proved the following results.

Theorem C ([4]). *Let f and g be two transcendental meromorphic functions, $n \geq 12$ an integer. If $f^n(f-1)f'$ and $g^n(g-1)g'$ share z CM, then either $f(z) \equiv g(z)$ or*

$$f = \frac{(n+2)h(1-h^{n+1})}{(n+1)(1-h^{n+2})}, \qquad g = \frac{(n+2)(1-h^{n+1})}{(n+1)(1-h^{n+2})}.$$

Theorem D ([4]). *Let f and g be two transcendental meromorphic functions, $n \geq 13$ an integer. If $f^n(f-1)^2 f'$ and $g^n(g-1)^2 g'$ share z CM, then $f \equiv g$.*

Now the following problem is considered: Is it possible to relax the nature of sharing values in the above theorems and if possible, how far? To answer this question the notion of weighted sharing of values is introduced as follows.

Definition 1. *Let k be a non-negative integer or infinity. For a complex number $a \in C \cup \{\infty\}$, we denote by $E_k(a,f)$ the set of all a-points of f where an a-point with multiplicity m is counted m times if $m \leq k$ and $k+1$ times if $m > k$.*

Definition 2. *Let k be a non-negative integer or infinity. For a complex number $a \in C \cup \{\infty\}$, such that $E_k(a,f) = E_k(a,g)$, then we say that f and g share the value a with weight k.*

The definition implies that if f, g share value a with weight k then z_0 is a zero of $f - a$ with multiplicity $m(\leq k)$ if and only if it is a zero of $g - a$ with multiplicity $m(\leq k)$ and $z_0$is a zero of $f - a$ with multiplicity $m(> k)$ if and only if it is a zero of $g - a$ with multiplicity $n(> k)$, where m is not necessarily equal to n.

We write f, g share (a,k) to mean that f, g share the value a with weight k. Clearly if f, g share (a,k) then f, g share (a,p) for all integer $p, 0 \leq p < k$. Also we note that f, g share a value a IM or CM if and only if f, g share $(a,0)$ or (a,∞), respectively.

Definition 3. *We say that f and g share (z,k) if $f - z$ and $g - z$ share $(0,k)$.*

Using the idea of weighted sharing of values, we prove the following theorems which are improvements of the ones above.

Theorem 1.1. *Let f and g be two transcendental meromorphic functions, $n \geq 12$ an integer. If $f^n(f-1)f'$ and $g^n(g-1)g'$ share $(z,2)$, then either $f(z) \equiv g(z)$ or*

$$f = \frac{(n+2)h(1-h^{n+1})}{(n+1)(1-h^{n+2})}, \qquad g = \frac{(n+2)(1-h^{n+1})}{(n+1)(1-h^{n+2})}$$

Theorem 1.2. *Let f and g be two transcendental meromorphic functions, $n \geq 13$ an integer. If $f^n(f-1)^2f'$ and $g^n(g-1)^2g'$ share $(z,2)$, then $f \equiv g$.*

2. Some Lemmas

Lemma 2.1 ([6]). *Let f be a nonconstant meromorphic function, and let $a_1, a_2, \ldots, a_n$ be finite complex numbers , $a_n \neq 0$. Then*

$$T(r, a_nf^n + \cdots + a_2f^2 + a_1f + a_0) = nT(r,f) + S(r,f).$$

Lemma 2.2 ([5]). *Let f and g share $(1,2)$. Then one of the following cases hold:*

$$T(r,f) \leq N_2\left(r, \frac{1}{f}\right) + N_2\left(r, \frac{1}{g}\right) + N_2(r,f) + N_2(r,g) + S(r,f) + S(r,g). \tag{1}$$

The same inequality holds for $T(r,g)$.

$$f \equiv g; \tag{2}$$
$$fg \equiv 1. \tag{3}$$

Lemma 2.3 ([7]). *Let*

$$Q(\omega) = (n-1)^2(\omega^n - 1)(\omega^{n-2} - 1) - n(n-2)(\omega^{n-1} - 1)^2,$$

then

$$Q(\omega) = (\omega - 1)^4(\omega - \beta_1)(\omega - \beta_2)\cdots(\omega - \beta_{2n-6}),$$

where $\beta_j \in C - \{0,1\}$ $(j = 1,2,\ldots,2n-6)$, which are distinct respectively.

Lemma 2.4. *Let f and g be two nonconstant meromorphic functions, $m < 7$, $n > 7 - m$ positive integers, $\alpha(z)$ denoted as in Section 1. and $\alpha \not\equiv 0, \infty$, and let*

$$F = f^n(f-1)^m f', \qquad G = g^n(g-1)^m g'.$$

If F and G share $(\alpha(z),2)$, then $S(r,f) = S(r,g)$.

Proof. By Lemma 2.1, we have

$$\begin{aligned}(n+m)T(r,f) &= T(r, f^n(f-1)^m) + S(r,f) \\ &\leq T(r,F) + T(r,f') + S(r,f),\end{aligned}$$

therefore,

$$T(r,F) \geq (n+m-2)T(r,f) + S(r,f).$$

By the second fundamental theory, we have

$$\begin{aligned} T(r,F) &\le \overline{N}(r,F)+\overline{N}\left(r,\frac{1}{F}\right)+\overline{N}\left(r,\frac{1}{F-\alpha}\right)+S(r,f) \\ &\le \overline{N}(r,f)+\overline{N}\left(r,\frac{1}{f}\right)+\overline{N}\left(r,\frac{1}{f-1}\right) \\ &\quad +\overline{N}\left(r,\frac{1}{f'}\right)+\overline{N}\left(r,\frac{1}{G-\alpha}\right)+S(r,f) \\ &\le 5T(r,f)+T(r,G)+S(r,f). \end{aligned}$$

Noting that

$$T(r,G)\le T(r,g^n(g-1)^m)+T(r,g')\le (n+m+2)T(r,g)+S(r,g),$$

we deduce

$$(n+m-7)T(r,f)\le (n+m+2)T(r,g)+S(r,g)+S(r,f).$$

It follows that the result holds. □

3. Proof of the Main Theorems

Proof (Proof of Theorem 1.1). Lemma 2.4 implies that $S(r,f)=S(r,g)$. Let

$$F=\frac{f^n(f-1)f'}{z},\qquad G=\frac{g^n(g-1)g'}{z}$$

and

$$F^*=\frac{1}{n+2}f^{n+2}-\frac{1}{n+1}f^{n+1},\qquad G^*=\frac{1}{n+2}g^{n+2}-\frac{1}{n+1}g^{n+1}.$$

Thus we obtain that F and G share $(1,2)$. Moreover, by Lemma 2.1, we have

$$\begin{aligned} T(r,F^*)&=(n+2)T(r,f)+S(r,f),\\ T(r,G^*)&=(n+2)T(r,g)+S(r,g). \end{aligned}$$

Since $(F^*)'=Fz$, we deduce

$$m\left(r,\frac{1}{F^*}\right)\le m\left(r,\frac{1}{zF}\right)+S(r,f)\le m\left(r,\frac{1}{F}\right)+\log r+S(r,f)$$

and by the first fundamental theorem

$$T(r,F^*)\le T(r,F)+N\left(r,\frac{1}{F^*}\right)-N\left(r,\frac{1}{F}\right)+\log r+S(r,f).$$

Note that

$$N\left(r,\frac{1}{F^*}\right)=(n+1)N\left(r,\frac{1}{f}\right)+N\left(r,\frac{1}{f-\frac{n+2}{n+1}}\right),$$

$$N\left(r,\frac{1}{F}\right)=nN\left(r,\frac{1}{f}\right)+N\left(r,\frac{1}{f-1}\right)+N\left(r,\frac{1}{f'}\right).$$

So we have

$$T(r,F^*)\leq T(r,F)+N\left(r,\frac{1}{f}\right)+N\left(r,\frac{1}{f-\frac{n+2}{n+1}}\right)$$
$$+\log r-N\left(r,\frac{1}{f-1}\right)-N\left(r,\frac{1}{f'}\right)+S(r,f).$$

Since F and G share $(1,2)$, by Lemma 2.2 we have one of the following cases hold:

(1) $T(r,F)\leq N_2\left(r,\frac{1}{F}\right)+N_2\left(r,\frac{1}{G}\right)+N_2(r,F)+N_2(r,G)+S(r,F)+S(r,G)$, the same inequality holds for $T(r,G)$.

(2) $FG\equiv 1$

(3) $F\equiv G$.

Next we consider three cases:

Case 5. (1) holds. So

$$T(r,F)\leq N_2\left(r,\frac{1}{F}\right)+N_2\left(r,\frac{1}{G}\right)+N_2(r,F)+N_2(r,G)+S(r,F)+S(r,G).$$

Obviously we have

$$N_2(r,F)+N_2\left(r,\frac{1}{F}\right)\leq 2\overline{N}(r,f)+2N\left(r,\frac{1}{f}\right)+N\left(r,\frac{1}{f-1}\right)+N\left(r,\frac{1}{f'}\right)+\log r$$

and

$$N_2(r,G)+N_2\left(r,\frac{1}{G}\right)\leq 2\overline{N}(r,g)+2N\left(r,\frac{1}{g}\right)+N\left(r,\frac{1}{g-1}\right)+N\left(r,\frac{1}{g'}\right)+\log r.$$

Thus

$$T(r,F^*)\leq 3N\left(r,\frac{1}{f}\right)+N\left(r,\frac{1}{f-\frac{n+2}{n+1}}\right)$$
$$+2N(r,f)+2N(r,g)+2N\left(r,\frac{1}{g}\right)+N\left(r,\frac{1}{g-1}\right)+N\left(r,\frac{1}{g'}\right)+3\log r+S(r,f).$$

Note that

$$N\left(r,\frac{1}{g'}\right)\leq N(r,g)+N\left(r,\frac{1}{g}\right)\leq 2T(r,g)+S(r,g),$$

we have

$$(n-4)T(r,f)\leq 7T(r,g)+3\log r+S(r,g).$$

In the same manner as above, we have

$$(n-4)T(r,g)\leq 7T(r,f)+3\log r+S(r,g).$$

We obtain that $n\leq 11$, which contradicts $n\geq 12$.

Case 6. *(2) holds. So* $FG \equiv 1$, *that is*

$$f^n(f-1)f'g^n(g-1)g' \equiv z^2.$$

Let $z_0(\neq 0,\infty)$ *be a zero of* f *of order* p, *so* z_0 *is a pole of* g. *Suppose that* z_0 *is pole of* g *of order* q. *From above we obtain*

$$np+p-1 = nq+2q+1$$

that is, $(n+1)(p-q) = q+2$, *which implies that* $p \geq q+1$ *and* $q+2 \geq n+1$. *Hence,* $p \geq n$.

Let $z_1(\neq 0,\infty)$ *be a zero of* $f-1$ *of order* p_1, *then we can also deduce that* $p_1 \geq \frac{n}{2}+2$.

Let $z_2(\neq 0,\infty)$ *be a zero of* f' *of order* p_2 *that is not a zero of* $f(f-1)$, *we similarly have* $p_2 \geq n+3$. *Moreover, in the same manner as above, we have similar results for the zeros of* $g^n(g-1)g'$. *On the other hand, suppose that* z_3 *is a pole of* f, *we get that* z_3 *is the zero of* $g^n(g-1)g'$, *thus*

$$\begin{aligned}\overline{N}(r,f) &\leq \overline{N}\left(r,\frac{1}{g}\right)+\overline{N}\left(r,\frac{1}{g-1}\right)+\overline{N}\left(r,\frac{1}{g'}\right)\\ &\leq \frac{1}{12}N\left(r,\frac{1}{g}\right)+\frac{1}{8}N\left(r,\frac{1}{g-1}\right)+\frac{2}{15}N\left(r,\frac{1}{g'}\right)\\ &< \frac{2}{3}T(r,g)+S(r,g).\end{aligned}$$

By the second fundamental theorem, we have

$$\begin{aligned}T(r,f) &\leq \overline{N}\left(r,\frac{1}{f}\right)+\overline{N}\left(r,\frac{1}{f-1}\right)+\overline{N}(r,f)+S(r,f)\\ &< \frac{5}{24}T(r,f)+\frac{2}{3}T(r,g)+2\log r+S(r,g).\end{aligned}$$

Similarly, we have

$$T(r,g) < \frac{5}{24}T(r,g)+\frac{2}{3}T(r,f)+2\log r+S(r,g).$$

From above we deduce a contradiction.

Case 7. *(3) holds. So* $F \equiv G$. *That is*

$$F^* \equiv G^*+c,$$

where c *is a constant. It follows that*

$$T(r,f) = T(r,g)+S(r,f).$$

Suppose that $c \neq 0$. By the second fundamental theorem, we have

$$\begin{aligned}(n+2)T(r,g) &= T(r,G^*)\\ &< \overline{N}\left(r,\frac{1}{G^*}\right)+\overline{N}\left(r,\frac{1}{G^*+c}\right)+\overline{N}(r,G^*)+S(r,g)\\ &\leq \overline{N}\left(r,\frac{1}{g}\right)+\overline{N}\left(r,\frac{1}{g-\frac{n+2}{n+1}}\right)+\overline{N}(r,g)\\ &\quad+\overline{N}\left(r,\frac{1}{f}\right)+\overline{N}\left(r,\frac{1}{f-\frac{n+2}{n+1}}\right)+S(r,f)\\ &\leq 5T(r,f)+S(r,f),\end{aligned}$$

which contradicts the assumption. Therefore $F^* \equiv G^*$, that is

$$f^{n+1}\left(f-\frac{n+2}{n+1}\right)=g^{n+1}\left(g-\frac{n+2}{n+1}\right).$$

Let $h=f/g$. If $h\equiv 1$, then $f\equiv g$. If $h\not\equiv 1$, we deduce that

$$f=\frac{(n+2)h(1-h^{n+1})}{(n+1)(1-h^{n+2})},\qquad g=\frac{(n+2)(1-h^{n+1})}{(n+1)(1-h^{n+2})},$$

where h is a nonconstant meromorphic function.

□

Proof (Proof of Theorem 1.2). Let

$$F=\frac{f^n(f-1)^2f'}{z},\qquad G=\frac{g^n(g-1)^2g'}{z}$$

and

$$F^*=\frac{1}{n+3}f^{n+3}-\frac{2}{n+2}f^{n+2}+\frac{1}{n+1}f^{n+1},$$
$$G^*=\frac{1}{n+3}g^{n+3}-\frac{2}{n+2}g^{n+2}+\frac{1}{n+1}g^{n+1}.$$

Thus we obtain that F and G share $(1,2)$. Moreover by Lemma 2.1, we have

$$T(r,F^*)=(n+3)T(r,f)+S(r,f),$$
$$T(r,G^*)=(n+3)T(r,g)+S(r,g).$$

Proceeding as in the proof of Theorem 1.1, we have

$$T(r,F^*)\leq T(r,F)+N\left(r,\frac{1}{F^*}\right)-N\left(r,\frac{1}{F}\right)+\log r+S(r,f).$$

Note that

$$N\left(r,\frac{1}{F^*}\right)=(n+1)N\left(r,\frac{1}{f}\right)+N\left(r,\frac{1}{f-a_1}\right)+N\left(r,\frac{1}{f-a_2}\right),$$

where a_1, a_2 are distinct roots of the algebraic equation $\frac{1}{n+3}z^2 - \frac{2}{n+2}z + \frac{1}{n+1} = 0$, and

$$N\left(r, \frac{1}{F}\right) = nN\left(r, \frac{1}{f}\right) + 2N\left(r, \frac{1}{f-1}\right) + N\left(r, \frac{1}{f'}\right).$$

Since F and G share $(1,2)$, by Lemma 2.2 we have one of the following cases hold:

(1) $T(r,F) \le N_2\left(r, \frac{1}{F}\right) + N_2\left(r, \frac{1}{G}\right) + N_2(r,F) + N_2(r,G) + S(r,F) + S(r,G)$, the same inequality holds for $T(r,G)$.

(2) $FG \equiv 1$;

(3) $F \equiv G$.

Next we consider three cases:

Case 8. *(1) holds. So*

$$T(r,F) \le N_2\left(r, \frac{1}{F}\right) + N_2\left(r, \frac{1}{G}\right) + N_2(r,F) + N_2(r,G) + S(r,F) + S(r,G)$$

Obviously we have

$$N_2(r,F) + N_2\left(r, \frac{1}{F}\right) \le 2\overline{N}(r,f) + 2N\left(r, \frac{1}{f}\right) + 2N\left(r, \frac{1}{f-1}\right) + N\left(r, \frac{1}{f'}\right) + \log r$$

and

$$N_2(r,G) + N_2\left(r, \frac{1}{G}\right) \le 2\overline{N}(r,g) + 2N\left(r, \frac{1}{g}\right) + 2N\left(r, \frac{1}{g-1}\right) + N\left(r, \frac{1}{g'}\right) + \log r.$$

Thus

$$\begin{aligned} T(r,F^*) \le 3N\left(r, \frac{1}{f}\right) + N\left(r, \frac{1}{f-a_1}\right) + N\left(r, \frac{1}{f-a_2}\right) + 2N(r,f) + 2N(r,g) \\ + 2N\left(r, \frac{1}{g}\right) + N\left(r, \frac{1}{g-1}\right) + N\left(r, \frac{1}{g'}\right) + 3\log r + S(r,f). \end{aligned}$$

Note that

$$N\left(r, \frac{1}{g'}\right) \le N(r,g) + N\left(r, \frac{1}{g}\right) \le 2T(r,g) + S(r,g),$$

we have

$$(n-4)T(r,f) \le 8T(r,g) + 3\log r + S(r,g).$$

In the same manner as above, we have

$$(n-4)T(r,g) \le 8T(r,f) + 3\log r + S(r,g).$$

we obtain a contradiction.

Case 9. *(2) holds. So* $FG \equiv 1$. *In the same manner as in the proof of Theorem 1.1, we again deduce a contradiction.*

Case 10. *(3) holds. So* $F \equiv G$. *Thus* $F^* \equiv G^*$, *that is*

$$\frac{1}{n+3}f^{n+3} - \frac{2}{n+2}f^{n+2} + \frac{1}{n+1}f^{n+1} = \frac{1}{n+3}g^{n+3} - \frac{2}{n+2}g^{n+2} + \frac{1}{n+1}g^{n+1}$$

set $h = f/g$, *we substitute* $f = hg$ *in the above, it follows that*

$$(n+2)(n+1)g^2(h^{n+3}-1) - 2(n+3)(n+1)g(h^{n+2}-1) \\ + (n+2)(n+3)(h^{n+1}-1) = 0$$

If h *is not a constant, using Lemma 2.3, we can conclude that*

$$\{(n+1)(n+2)(h^{n+3}-1)g - (n+3)(n+1)g(h^{n+2}-1)\}^2 = -(n+3)(n+1)Q(h)$$

where

$$Q(h) = (h-1)^4(h-\beta_1)(h-\beta_2)\cdots(h-\beta_{2n}), \quad \beta_j \in C - \{0,1\}\ (j = 1,2,\ldots,2n),$$

which are pairwise distinct. This implies that every zero of $h - \beta_j (j = 1,2,\ldots,2n)$ *has a multiplicity of at least 2. By the second fundamental theorem we obtain that* $n \leq 2$, *which is again a contradiction. Therefore* h *is a constant. We have that* $h^{n+1} - 1 = 0$ *and* $h^{n+2} - 1 = 0$, *which implies that* $h = 1$, *and hence* $f \equiv g$. *This completes the proof of the theorem.*

□

Bibliography

[1] W.K. Hayman, *Meromorphic Functions*, Clarendon, Oxford, 1964.

[2] C.C. Yang and X.H. Hua, Uniqueness and value-sharing of meromorphic functions, *Ann. Acad. Sci. Fenn. Math.*, **22** (1997), 395–406.

[3] M.L. Fang and W. Hong, A unicity theorem for entire functions concerning differential polynimials, *Indian J. Pure Appl. Math.*, **32** (2001), 1343–1348.

[4] W.C. Lin and H.X. Yi, Uniqueness theorems for meromorphic functions concerning fixed-points, *Complex Variables Theory and Applications*, **49** (2004), 793–806.

[5] I. Lahiri, Weighted value sharing and uniqueness of meromorphic functions, *Complex Variables Theory and Applications*, **46** (2001), 241–253.

[6] C.C. Yang, On deficiencies of differential polynomials, *Math. Z.*, **125** (1972), 107–112.

[7] G. Frank and M. Reiders, A unique range set for meromorphic functions with 11 elements, *Complex Variables Theory and Applications*, **37** (1998), 185–193.

In: Advances in Inequalities for Series ISBN 978-1-60021-920-7
Editors: S. S. Dragomir and A. Sofo, pp. 181–187

Chapter 12

Some Conditions of Boundedness for a Class of Spiral Like Integral Operators

Gheorghe Miclăuş
Department of Mathematics and Computer Science
Victoriei 76, 4800 Baia Mare, Romania
E-mail address: miclaus5@yahoo.com

Abstract

In this paper we determine the Hardy spaces of the class of integral operators defined by (1.2) and their iterative, when they are applied to the class of spirallike functions. As a corollary we obtain some conditions for their boundedness.

1991 Mathematics Subject Classification: 30C45, 30D55

Key words and phrases: Holomorphic functions, integral operators, Hardy spaces.

1. Introduction

Let A denote the set of functions $f(z) = z + a_2 z^2 + \cdots$ that are analytic in the unit disc U, and let S denote the subset of A consisting of univalent functions. Let α and ρ be real numbers satisfying $|\alpha| < \frac{\pi}{2}$ and $\rho < 1$. A function $f \in A$ is said to belong to the class $\widehat{S}_\rho(\alpha)$ if

$$\operatorname{Re}\left[e^{i\alpha}\frac{zf'(z)}{f(z)}\right] > \rho\cos\alpha, \quad \text{for} \quad z \in U \tag{1.1}$$

Note that $\widehat{S}_0(0) = S^*$, the class of starlike functions.

If $0 \le \rho < 1$, then $\widehat{S}_\rho(\alpha)$ is the well-known class of univalent functions known as the $\alpha-$spirallike functions of order ρ. In particular, when $\rho = 0$ then $\widehat{S}_0(\alpha)$ is the class of $\alpha-$spirallike or spirallike functions.

The class of spirallike functions was introduced by L. Špaček in 1932 [8].

In 1970 T. Başgöze and F.R. Keogh [2] determined the Hardy classes of spirallike functions and its derivative. During the last several years many authors have employed various methods to study different types of integral operators $I(f)$ mapping subsets of S into S.

In 1973, R. Singh [7] showed that if

$$I(f)(z) = I_{\beta,\gamma}(f)(z) = \left[\frac{\beta+\gamma}{z^{\gamma}}\int_0^z f^{\beta}(t)t^{\gamma-1}dt\right]^{\frac{1}{\beta}} \tag{1.2}$$

and if $\beta,\gamma = 1,2,3,\ldots$then $I(S^*) \subset S^*$.

In 1985, S.S. Miller and P.T. Mocanu [6] determined conditions for which $I(\widehat{S}_{\rho}(\alpha)) \subset \widehat{S}_{\rho}(\alpha)$.

In 1997 [5] we have determined the Hardy spaces of the Singh integral operator (1.2) when f belongs to H^p.

In this paper we determine the Hardy spaces of the Singh integral operator (1.2) and its iterative when f is a spirallike function and as a corollary we obtain some conditions for their boundedness.

Also, in particular, we obtain conditions for boundedness of the Alexander integral operator [1] defined by (3.4) and the Libera generalized integral operator [3] defined by (3.5) when they are applied to spirallike functions.

2. Preliminaries

For f analytic in U and $z = re^{i\theta}$, we denote

$$M_p(r,f) = \begin{cases} \left(\frac{1}{2\pi}\int_0^{2\pi}\left|f(re^{i\theta})\right|^p d\theta\right)^{\frac{1}{p}}, & 0 < p < \infty, \\ \sup\limits_{0\le p\le 2\pi}\left|f(re^{i\theta})\right|, & \text{for } p = \infty. \end{cases}$$

A function f analytic in U is said to be a Hardy space $H^p, 0 < p \le \infty$, if $M_p(r,f)$ remains bounded as $r \to 1^-$. H^{∞} is the space of bounded analytic functions in the unit disc.

We shall need the following lemmas to prove our results.

Lemma 2.1 ([2]). *If $f \in \widehat{S}_{\rho}(\alpha)$ and f is not of the form*

$$\frac{z}{(1-ze^{i\tau})^a}, \qquad a = 2(1-\rho)(\cos\alpha - \sin\alpha)\cos\alpha, \quad \tau \in \mathbb{R}, \tag{2.1}$$

then there exists $\varepsilon = \varepsilon(f) > 0$ such that

$$f \in H^p, \quad p = \frac{1}{2(1-\rho)\cos^2\alpha} + \varepsilon.$$

Lemma 2.2 ([6]). *Let β and γ be complex numbers with* $\operatorname{Re}\beta > 0$ *and suppose ρ satisfies*

$$\frac{-\operatorname{Re}\gamma}{\operatorname{Re}\beta} \le \rho < 1.$$

If $\beta = |\beta| e^{i\alpha}$ and $f \in \widehat{S}_\rho(\alpha)$ then the Singh integral operator $I(f)$ defined by (1.2) is analytic and

$$I(f) \in \widehat{S}_\rho(\alpha).$$

Lemma 2.3 ([5]). *Let f be analytic in $U, f \in H^p$, and I is the Singh integral operator defined by (1.2) with $\beta > 0$, then:*

(i) if $\beta > np$ then $I^n(f) \in H^{\frac{\beta p}{\beta - np}}$

(ii) if $\beta \leq np$ then $I^n(f) \in H^\infty$ where $I^n = \underbrace{I \circ I \circ \ldots \circ I}_{n}$, $n \in \mathbb{N}^*$

3. Main Results

Theorem 3.1. *Let $\alpha, \beta, \gamma, \rho$ be real numbers with $|\alpha| < \frac{\pi}{2}$, $\beta > 0$, $0 \leq \rho < 1$. If $I \in \widehat{S}_\alpha(\rho)$ and I is the Singh integral operator defined by (1.2), then there exists $\varepsilon = \varepsilon(f) > 0$ such that:*

(i) if $\beta > \frac{1}{2(1-\rho)\cos^2\alpha} + \varepsilon$, then $I(f) \in H^{\beta p}$, where

$$p = \frac{\beta + 2\varepsilon(1-\rho)\cos^2\alpha}{(\beta - \varepsilon)2(1-\rho)\cos^2\alpha - 1};$$

(ii) if $\beta \leq \frac{1}{2(1-\rho)\cos^2\alpha} + \varepsilon$, then $I(f) \in H^\infty$.

Proof. From Lemma 2.3, for $n = 1$, we have that if $f \in H^p$ then

$$I(f) \in H^{\frac{\beta p}{\beta - p}}, \quad \text{for} \quad \beta > p \tag{3.1}$$

and

$$I(f) \in H^\infty, \quad \text{for} \quad \beta \leq p. \tag{3.2}$$

From Lemma 2.1, we have that if $f \in \widehat{S}_\rho(\alpha)$ and f is not of the form (2.1) then there exists $\varepsilon = \varepsilon(f) > 0$ such that

$$f \in H^p, \quad \text{where} \quad p = \frac{1}{2(1-\rho)\cos^2\alpha}. \tag{3.3}$$

From (3.1) and (3.3) we obtain assertion (i) and from (3.2) and (3.3) we obtain assertion (ii). □

Theorem 3.2. *Let $\alpha, \beta, \gamma, \rho$ be real numbers with $|\alpha| < \frac{\pi}{2}$, $\beta > 0$, $0 \leq \rho < 1$. If $f \in \widehat{S}_\rho(\alpha)$, f is not of the form (2.1) and I is the Singh integral operator defined by (1.2) then there exists $\varepsilon = \varepsilon(f) > 0$ such that*

(i) if $\beta > \frac{n}{2(1-\rho)\cos^2\alpha} + \varepsilon$, then $I^n(f) \in H^{\beta p}$, where

$$p = \frac{\beta + 2\varepsilon(1-\rho)\cos^2\alpha}{(\beta - n\varepsilon)2(1-\rho)\cos^2\alpha - n}, \qquad n \in \mathbb{N}^*;$$

(ii) if $\beta \leq \frac{n}{2(1-\rho)\cos^2\alpha} + n\varepsilon$, *then* $I^n(f) \in H^\infty$, $n \in \mathbb{N}^*$.

Proof. From Lemma 2.1 we have that if $f \in \widehat{S}_\rho(\alpha)$ and is not of the form (2.1) then there exists $\varepsilon = \varepsilon(f) > 0$ such that

$$f \in H^p, \quad p = \frac{1}{2(1-\rho)\cos^2\alpha}$$

From Lemma 2.3 we obtain the results. □

Using these results we obtain some boundedness conditions for the Singh integral operator and its iterative.

Corollary 3.3. *Let* α, β, ρ *be real numbers satisfying the conditions:*

$$|\alpha| < \frac{\pi}{2}, \quad 0 < \beta \leq \frac{1}{2(1-\rho)\cos^2\alpha},$$

$0 \leq \rho < 1$ *and* $\gamma \in \mathbb{C}$ *with*

$$-\frac{\operatorname{Re}\gamma}{\beta} \leq \rho.$$

If $f \in \widehat{S}_\rho(\alpha)$ *and* f *is not of the form (2.1) and if* I *is the integral operator defined by (1.2) then* $I(f)$ *is bounded.*

Indeed, from Theorem 3.1 and Lemma 2.2 we obtain the result.

Corollary 3.4. *Let* α, β, ρ *be real numbers satisfying the conditions:*

$$|\alpha| < \frac{\pi}{2}, \quad 0 < \beta \leq \frac{n}{2(1-\rho)\cos^2\alpha}, \quad 0 \leq \rho < 1$$

and $\gamma \in \mathbb{C}$ *with*

$$\frac{-\operatorname{Re}\gamma}{\beta} \leq \rho.$$

If $f \in \widehat{S}_\rho(\alpha)$, f *is not of the form (2.1) and if* I *is the Singh integral operator (1.2) then*

$I^n(f)$ *is bounded for all* $n \in \mathbb{N}^*$.

In particular, we obtain similarly conditions for the Alexander integral operator and for the Libera generalized integral operator. If in (1.2) we set $\beta = 1$ and $\gamma = 0$ then we obtain the Alexander integral operator:

$$I_{1,0}(f)(z) = \int_0^z f(t)t^{-1}dt. \tag{3.4}$$

Corollary 3.5. *Let* α, ρ *be real numbers with* $|\alpha| < \frac{\pi}{2}$, $0 \leq \rho < 1$. *If* $f \in \widehat{S}_\rho(\alpha)$, f *not of the form (2.1) and if* $I_{1,0}$ *is the Alexander integral operator (3.4) and* $2(1-\rho)\cos^2\alpha \leq 1$ *then:*

$I_{1,0}(f)$ *is bounded.*

If in (1.2) we set $\beta = 1$ then we obtain the Libera generalized integral operator:

$$I_{1,\gamma}(f)(z) = \frac{1+\gamma}{z^\gamma} \int_0^z f(t) t^{\gamma-1} dt \tag{3.5}$$

For this operator we have the following result.

Corollary 3.6. *Let α, ρ be real numbers with*

$$|\alpha| < \frac{\pi}{2}, \quad 0 \leq \rho < 1,$$

and $\gamma \in \mathbb{C}$ with $-\mathrm{Re}\gamma \leq \rho$.

If $f \in \widehat{S}_\rho(\alpha)$, f is not of the form (2.1) and if $I_{1,\gamma}$ is the Libera generalized integral operator (Bernardi) and if

$$2(1-\rho)\cos^2\alpha \leq 1, \quad \text{then } I_{1,\gamma}(f) \text{ is bounded.}$$

If we set $\rho = 0$ then $\widehat{S}_0(\alpha)$ is the class of spirallike functions. From Corollary 3.3 we obtain that if $\beta \leq \frac{1}{2}$ then the Singh integral operator maps the class $\widehat{S}_0(\alpha)$ into the subclass of the bounded functions.

Also, from Corollary 3.4 we have that the Singh integral operator mapping any function $f \in \widehat{S}_\rho(\alpha)$ into the bounded function after a number of iteratives.

Bibliography

[1] J.W. Alexander, Functions which map the interior of the unit circle upon simple regions, *Ann. of Math.*, **17** (1915), 12–22.

[2] T. Başgőze and F.R. Keogh, The Hardy class of a spirallike function and its derivative, *Proc. Amer. Math. Soc.*, **26** (1970), 266–269.

[3] S.D. Bernardi, Convex and starlike univalent functions, *Trans. Amer. Math. Soc.*, **135** (1969), 266–269.

[4] P.L. Duren, *Theory of H^p Spaces*, Academic Press, New York and London, 1970.

[5] Gh. Miclăuş, Integral operator of Singh and Hardy classes, *Studia Univ. "Babeş-Bolyai" Mathematica*, **42**(2) (1997), 71–77.

[6] S.S. Miller and P.T. Mocanu, On a class of spirallike integral operators, *Rev. Roumaine Math. Pures Appl.*, **31** (1986), 225–230.

[7] R. Singh, On Bazilevic functions, *Proc. Amer. Math. Soc.*, **18** (1973), 225–230.

[8] L. Špaček, Contribution à la theorie des functions univalentes, *Casopis Pest. Mat.*, **62** (1932), 12–19.

In: Advances in Inequalities for Series
Editors: S. S. Dragomir and A. Sofo, pp. 189–197
ISBN 978-1-60021-920-7

Chapter 13

On Certain Classes of Analytic Functions of Complex Order

G. Murugusundaramoorthy[1] *and Thomas Rosy*[2]
[1]Department of Mathematics
Vellore Institute of Technology
Deemed University
Vellore - 632014, India
E-mail address: gmsmoorthy@yahoo.com
[2]Department of Mathematics
Madras Christian College
Chennai - 600059, India
E-mail address: drthomasrosy@rediffmail.com

Abstract

We introduce a new subclass of analytic functions making use of Salagean operator in the unit disc. A sufficient condition, maximization theorem and some inclusion relations associated with (n,δ) neighborhoods of the functions belonging to the class are obtained.

1991 Mathematics Subject Classification: 30C45

Key words and phrases: Univalent, analytic, convex, starlike, complex order, Salagean derivative.

1. Introduction and Preliminaries

Let S denote the class of functions f of the form

$$f(z) = z + \sum_{k=2}^{\infty} a_k z^k \tag{1.1}$$

which are analytic and univalent in the unit disc $U = \{z : z \in \mathcal{C} \text{ and } |z| < 1\}$. Let Ω denote the class of functions $w(z)$ in U satisfying the condition $w(0) = 0$ and $w(z) < 1$ for $z \in U$.

For a function $f(z) \in S$, we define due to Salagean [7],

$$D^0 f(z) = f(z)$$
$$D^1 f(z) = Df(z) = zf'(z)$$
$$\text{and} \quad D^n f(z) = D(D^{n-1} f(z)), \qquad n \in \mathbb{N} = \{1, 2 \ldots\}$$

with the help of the differential operator D^n, we say that the function $f(z) \in S$ is in the class $S(\lambda, \beta, \gamma, n)$ if and only if

$$\left| \frac{\frac{zF'(z)}{F(z)} - 1}{2\beta\gamma + \frac{zF'(z)}{F(z)} - 1} \right| < 1, \qquad z \in U, \tag{1.2}$$

where $0 < \beta \leq 1$, $\gamma \in C \backslash \{0\}$ and

$$\frac{zF'(z)}{F(z)} = \frac{(1-\lambda)z(D^n f(z))' + \lambda z(D^{n+1} f(z))'}{(1-\lambda)D^n f(z) + \lambda(D^{n+1} f(z))} \tag{1.3}$$

$0 \leq \lambda \leq 1$ and $n \in N \cup \{0\}$.

Denote by T, the subclass of S consisting of functions of the form

$$f(z) = z - \sum_{k=2}^{\infty} a_k z^k, \qquad a_k \geq 0. \tag{1.4}$$

Finally, let $ST(\lambda, \beta, \gamma, n)$ denote the subclass of T consisting of functions $f(z)$ of the form (1.4) which satisfy the inequality

$$\left| \frac{1}{\gamma} \left(\frac{zF'(z)}{F(z)} - 1 \right) \right| < \beta \tag{1.5}$$

also, let $RT(\lambda, \beta, \gamma, n)$ denote the subclass of S consisting of functions $f(z)$ of the form (1.4) which satisfy the inequality

$$\left| \frac{1}{\gamma} \{(1-\lambda)(D^n f(z))' + \lambda(D^{n+1} f(z))' - 1\} \right| < \beta \tag{1.6}$$

for $z \in U$, $\gamma \in C \backslash \{0\}$; $0 \leq \lambda \leq 1$, $0 < \beta \leq 1$. Various subclasses of the classes $ST(\lambda, \beta, \gamma, n)$ and $RT(\lambda, \beta, \gamma, n)$ with $\gamma = 1$ were studied in many earlier works [1, 2, 8, 9, 10, 11].

The purpose of the present paper is to determine sufficient conditions in terms of coefficient bounds for functions to belong to $S(\lambda, \beta, \gamma, n)$. Coefficient estimates and maximization of $|a_3 - \mu a_2^2|$ on the class $S(\lambda, \beta, \gamma, n)$ for complex values of μ are determined. Further we investigate $(n - \delta)$ neighborhoods of the classes $ST(\lambda, \beta, \gamma, n)$ and $RT(\lambda, \beta, \gamma, n)$ with negative coefficients.

2. A Sufficient Condition for a Function to be in $S(\lambda, \beta, \gamma, n)$

Theorem 2.1. *Let the function $f(z)$ be defined by (1.1) and let*

$$\sum_{k=2}^{\infty} k^n(\lambda k + 1 - \lambda)\{k + 2\beta|\gamma| - 1\} a_k \leq \beta|\gamma|, \tag{2.1}$$

hold then $f(z)$ belongs to $S(\lambda, \beta, \gamma, n)$.

Proof. Suppose that the inequality (2.1) holds. Then we have for $z \in U$:

$$|zF'(z) - F(z)| - |2\beta\gamma F(z) + (zF'(z) - F(z))|$$
$$= \left|\sum_{k=2}^{\infty} k^n (k-1)(\lambda k + 1 - \lambda) a_k\right| - \left|2\beta\gamma + \sum_{k=2}^{\infty} k^n (\lambda k + 1 - \lambda)(k - 1 + 2\beta\gamma) a_k\right|$$
$$= \sum_{k=2}^{\infty} k^n (\lambda k + 1 - \lambda)\{k + 2\beta|\gamma| - 1\} a_k - 2\beta|\gamma|.$$
$$\leq 0, \quad \text{by } (2.1).$$

Hence it follows that

$$\left| \frac{\frac{zF'(z)}{F(z)} - 1}{2\beta\gamma + \left(\frac{zF'(z)}{F(z)} - 1\right)} \right| < 1, \qquad z \in U.$$

Letting

$$w(z) = \frac{\frac{zF'(z)}{F(z)} - 1}{2\beta\gamma + \left(\frac{zF'(z)}{F(z)} - 1\right)},$$

then $w(0) = 0$, $w(z)$ is analytic in $|z| < 1$ and $|w(z)| < 1$. Hence we have

$$\frac{zF'(z)}{F(z)} = \frac{1 + (2\beta|\gamma| - 1)w(z)}{1 - w(z)}.$$

□

3. Maximization of $|a_3 - \mu a_2^2|$

We need the following lemma for our discussion:

Lemma 3.1 ([4]). *Let* $w(z) = \sum_{k=2}^{\infty} c_k z^k \in \Omega$. *If* μ *is any complex number, then* $|c_2 - \mu c_1^2| \leq \max\{1, |\mu|\}$.

Theorem 3.2. *If a function* $f(z)$ *defined by (1.1) is in the class* $S(\lambda, \beta, \gamma, n)$ *and* μ *is any complex number, then*

$$|a_3 - \mu a_2^2| \leq \max\{1, |d|\} \frac{\beta|\gamma|}{3^n(2\lambda + 1)},$$

where

$$d = \frac{3^n(2\lambda + 1)\mu\beta|\gamma|}{2^{2n-2}(1 + \lambda)^2} - [2\beta|\gamma| + 1].$$

Proof.

$$w(z) = \frac{zF'(z) - F(z)}{(2\beta|\gamma| - 1)F(z) + zF'(z)}$$
$$= \frac{\sum_{k=2}^{\infty} k^n (k-1)(\lambda k + 1 - \lambda) a_k z^{k-1}}{2\beta|\gamma| + \sum_{k=2}^{\infty} k^n (\lambda k + 1 - \lambda)[2\beta\gamma + (k-1)] a_k z^{k-1}}$$
$$= \frac{\sum_{k=2}^{\infty} k^n (k-1)(\lambda k + 1 - \lambda) a_k z^{k-1}}{2\beta|\gamma|} \left(\frac{1 + \sum_{k=2}^{\infty} k^n (\lambda k + 1 - \lambda)[2\beta\gamma + k - 1] a_k z^{k-1}}{2\beta|\gamma|} \right)^{-1}$$

That is,

$$w(z) \left[1 + \frac{1}{2\beta|\gamma|} \sum_{k=2}^{\infty} k^n (\lambda k + 1 - \lambda)[2\beta|\gamma| + (k-1)] \right] a_k z^{k-1}$$
$$= \frac{\sum_{k=2}^{\infty} k^n (k-1)(\lambda k + 1 - \lambda) a_k z^{k-1}}{2\beta|\gamma|}. \quad (3.1)$$

Now comparing the coefficients of z and z^2 on both sides of (3.1), we thus obtain

$$a_2 = \frac{\beta|\gamma|}{2^{n-1}(1+\lambda)} c_1$$

and

$$a_3 = \{c_2 + c_1^2(2\beta|\gamma| + 1)\} \frac{\beta|\gamma|}{3^n(2\lambda+1)}.$$

Hence

$$a_3 - \mu a_2^2 = \frac{\beta|\gamma|}{3^n(2\lambda+1)} \left[c_2 - d c_1^2 \right], \quad (3.2)$$

where

$$d = \frac{3^n(2\lambda+1)\mu\beta|\gamma|}{2^{2n-2}(1+\lambda)^2} - [2\beta|\gamma| + 1].$$

Taking the modulus on both sides in (3.2), we have

$$\left| a_3 - \mu a_2^2 \right| \leq \frac{\beta|\gamma|}{3^n(2\lambda+1)} \left| c_2 - d c_1^2 \right|. \quad (3.3)$$

Using the lemma in (3.3), we have

$$\left| a_3 - \mu a_2^2 \right| \leq \frac{\beta|\gamma|}{3^n(2\lambda+1)} \max\{1, |d|\}.$$

□

In the following theorems we obtain necessary and sufficient conditions for a function to be in $ST(\lambda, \beta, \gamma, n)$ and $RT(\lambda, \beta, \gamma, n)$.

Theorem 3.3. *Let the function $f(z)$ be defined by (1.4), then $f(z)$ is in the class $ST(\lambda, \beta, \gamma, n)$ if and only if*

$$\sum_{k=2}^{\infty} k^n \{\lambda(k-1) + 1\} [k + \beta|\gamma| - 1] a_k \leq \beta|\gamma|. \quad (3.4)$$

Proof. Suppose that $f \in S(\lambda, \beta, \gamma, n)$. Then by appealing to condition (1.6), we readily get

$$\operatorname{Re}\left\{\frac{zF'(z)}{F(z)} - 1\right\} > -\beta|\gamma|, \qquad z \in U, \tag{3.5}$$

or equivalently

$$\operatorname{Re}\left\{\frac{-\sum_{k=2}^{\infty} k^n[\lambda(k-1)+1](k-1)a_k z^k}{z - \sum_{k=2}^{\infty} k^n[\lambda(k-1)+1]a_k z^k}\right\} > -\beta|\gamma|, \tag{3.6}$$

where we have made use of definition (1.5).

Now by choosing values of z on the real axis and allowing $z \to 1$; the inequality (3.6) immediately yields the desired condition (3.4). Conversely by applying to the hypothesis (3.4) and letting $|z| = 1$, we find that

$$\begin{aligned}\left|\frac{zF'(z)}{F(z)} - 1\right| &= \left|\frac{\sum_{k=2}^{\infty} k^n[\lambda(k-1)+1](k-1)a_k z^k}{z - \sum_{k=2}^{\infty} k^n[\lambda(k+1)-1]a_k z^k}\right| \\ &\le \frac{\beta|\gamma|\,(1 - \sum_{k=2}^{\infty} k^n[\lambda(k-1)+1]a_k)}{1 - \sum_{k=2}^{\infty} k^n[\lambda(k-1)+1]a_k} \\ &\le \beta|\gamma|.\end{aligned}$$

Hence by the maximum modulus theorem, we have $f \in ST(\lambda, \beta, \gamma, n)$, which completes the proof of Theorem 3.3. □

Similarly, we can prove the following;

Theorem 3.4. *Let the function $f \in S$ be defined by (1.4), then $f(z)$ is in the class $RT(\lambda, \beta, \gamma, n)$ if and only if*

$$\sum_{k=2}^{\infty} k^{n+1}[\lambda(k-1)+1]a_k \le \beta|\gamma|. \tag{3.7}$$

Remark 3.1. *A special case of Theorem 3.3, when $\gamma = 1$and $\beta = 1 - \alpha$, $(0 \le \alpha < 1)$ was given earlier by [11].*

4. Inclusion Relations Involving $N_{n,\delta}(e)$.

To study the inclusion relations involving $N_{n,\delta}(e)$ we need the following definitions. Following [2, 3, 6], we define the n, δ neighborhood of function $f(z) \in T$ by

$$N_{n,\delta}(f) = \left\{g \in T : g(z) = z - \sum_{k=2}^{\infty} b_k z^k \quad \text{and} \quad \sum_{k=2}^{\infty} k|a_k - b_k| \le \delta\right\}. \tag{4.1}$$

In particular, for the identity function $e(z) = z$, we have

$$N_{n,\delta}(e) = \left\{g \in T : g(z) = z - \sum_{k=2}^{\infty} b_k z^k \quad \text{and} \quad \sum_{k=2}^{\infty} k|b_k| \le \delta\right\}. \tag{4.2}$$

Theorem 4.1. *Let*

$$\delta = \frac{\beta|\gamma|}{2^{n-1}(\lambda+1)(1+\beta|\gamma|)}, \tag{4.3}$$

then $ST(\lambda,\beta,\gamma,n) \subset N_{n,\delta}(e)$.

Proof. For $f \in ST(\lambda,\beta,\gamma,n)$, Theorem 3.3 yields

$$2^n(\lambda+1)(1+\beta|\gamma|)\sum_{k=2}^{\infty} a_k \leq \beta|\gamma|,$$

so that

$$\sum_{k=2}^{\infty} a_k \leq \frac{\beta|\gamma|,}{2^n(\lambda+1)(1+\beta|\gamma|)}. \tag{4.4}$$

On the other hand, from (3.4) and (4.4) we have

$$\begin{aligned}
2^n(\lambda+1)\sum_{k=2}^{\infty} ka_k &\leq \beta|\gamma| + (1-\beta|\gamma|)[2^n(\lambda+1)]\sum_{k=2}^{\infty} a_k \\
&\leq \beta|\gamma| + (1-\beta|\gamma|)(2^n(\lambda+1))\frac{\beta|\gamma|,}{2^n(\lambda+1)(1+\beta|\gamma|)} \\
&\leq \beta|\gamma| + \beta|\gamma|\frac{(1-\beta|\gamma|)}{1+\beta|\gamma|} \\
&\leq \beta|\gamma|\left[1+\frac{(1-\beta|\gamma|)}{1+\beta|\gamma|}\right] \\
&\leq \frac{2\beta|\gamma|}{1+\beta|\gamma|},
\end{aligned}$$

$$\sum_{k=2}^{\infty} ka_k \leq \frac{2\beta|\gamma|}{2^n(\lambda+1)(1+\beta|\gamma|)} = \frac{\beta|\gamma|}{2^{n-1}(\lambda+1)(1+\beta|\gamma|)}. \tag{4.5}$$

□

Theorem 4.2. *Let* $\delta = \frac{2\beta|\gamma|}{2^{n+1}(\lambda+1)}$, *then* $RT(\lambda,\beta,\gamma,n) \subset N_{n,\delta}(e)$.

Proof. For $f \in RT(\lambda,\beta,\gamma,n)$, Theorem 3.4, immediately yields

$$2^{n+1}(\lambda+1)\sum_{k=2}^{\infty} a_k \leq \beta|\gamma|,$$

so that

$$\sum_{k=2}^{\infty} a_k \leq \frac{\beta|\gamma|}{2^{n+1}(\lambda+1)}. \tag{4.6}$$

On the other hand, we also find from (3.7) and (4.6) that

$$\begin{aligned}
\lambda 2^{n+1}\sum_{k=2}^{\infty} ka_k &\leq \beta|\gamma| + (\lambda-1)2^{n+1}\sum_{k=2}^{\infty} a_k \\
&\leq \beta|\gamma| + (\lambda-1)2^{n+1}\frac{\beta|\gamma|}{2^{n+1}(\lambda+1)}
\end{aligned}$$

$$\sum_{k=2}^{\infty} ka_k \leq \frac{2\beta|\gamma|}{2^{n+1}(\lambda+1)}.$$

□

Remark 4.1. *A special case of Theorem 4.2, when* $\gamma = 1$, $\beta = 1-\alpha$ $(0 \leq \alpha < 1)$, $n = 0$, $\lambda = 0$ *was proved recently by Altintas and Owa [1].*

Now we determine the neighborhood for each of the classes $ST(\lambda,\beta,\gamma,n)$ and $RT(\lambda,\beta,\gamma,n)$ which we define as follows. A function $f \in T$ is said to be in the class $ST(\lambda,\beta,\gamma,n)$ if there exists a function $g \in ST(\lambda,\beta,\gamma,n)$ such that

$$\left|\frac{f(z)}{g(z)} - 1\right| < 1-\alpha, \qquad (z \in U,\ 0 \leq \alpha < 1). \tag{4.7}$$

Analogously, A function $f \in T$ is said to be in the class $RT(\lambda,\beta,\gamma,n)$ if there exists a function $g \in RT(\lambda,\beta,\gamma,n)$ such that the inequality (4.6) holds true.

Theorem 4.3. *If* $g \in ST(\lambda,\beta,\gamma,n)$ *and*

$$\alpha = 1 - \frac{2^{n-1}(1+2\lambda)(1+\beta|\gamma|)\delta}{[2^n(1+2\lambda)(1+\beta|\gamma|) - \beta|\gamma|]},$$

then $N_\delta(g) \subset ST(\alpha,\lambda,\beta,\gamma,n)$.

Proof. Suppose that $f \in N_\delta(g)$ then we find from (4.6) that

$$\sum_{k=2}^{\infty} |a_k - b_k| \leq \delta,$$

which implies that the coefficient inequality

$$\sum_{k=2}^{\infty} |a_k - b_k| \leq \frac{\delta}{2}.$$

Next, since $g \in \subset ST(\lambda,\beta,\gamma,n)$, we have

$$\sum_{k=2}^{\infty} b_k \leq \frac{\beta|\gamma|}{2^n(1+\lambda)(1+\beta|\gamma|)},$$

so that

$$\begin{aligned}\left|\frac{f(z)}{g(z)} - 1\right| &< \frac{\sum_{k=2}^{\infty}|a_k - b_k|}{1-\sum_{k=2}^{\infty} b_k}\\ &\leq \frac{\delta}{2} \times \frac{2^n(1+\lambda)(1+\beta|\gamma|)}{2^n(1+\lambda)(1+\beta|\gamma|) - \beta|\gamma|}\\ &\leq 1-\alpha,\end{aligned}$$

provided that α is given precisely by (4.6). Thus by definition, $f \in ST(\alpha,\lambda,\beta,\gamma,n)$ for α given by (4.6), which completes the proof. □

Theorem 4.4. *If* $g \in RT(\lambda,\beta,\gamma,n)$ *and*

$$\alpha = 1 - \frac{2^n(1+\lambda)\delta}{2^{n+1}(1+\lambda) - \beta|\gamma|},$$

then $N_\delta(g) \subset RT(\alpha,\lambda,\beta,\gamma,n)$.

Acknowledgement

The authors would like to thank Prof. Jay M. Jagangiri, Kent State University, Burton, Ohio, U.S.A, for his valuable suggestions and comments which led to the improvement of the results.

Bibliography

[1] O. Altintas and S. Owa, Neighborhood of certain analytic functions with negative coefficients, *International J. Math. and Math. Sci.*, **19** (1996), 797–800.

[2] O. Altintas, O. Ozkan and H.M. Srivastava, Neighborhoods of a class of analytic functions with negative coefficients, *Appl. Math. Lett.*, **13** (2000), 63–67.

[3] A.W. Goodman, Univalent functions and nonanalytic curves, *Proc. Amer. Math. Soc.*, **8** (1957), 598–601.

[4] F.R. Keogh and E.P. Merkes, A coefficient inequality for certain classes of analytic functions, *Proc. Amer. Math. Soc.*, **20** (1969), 8–12.

[5] M.A. Nasar and M.K. Aouf, Starlike functions of complex order, *J. Natur. Sci. Math.*, (1985), 1–12.

[6] S. Ruscheweyh, Neighborhoods of univalent functions, *Proc. Amer. Math. Soc.*, **81** (1981), 521–527.

[7] G.S. Salagean, Subclasses of univalent functions, *Lect. Notes in Maths.*(Springer - Verlag), 1013, (1983), 362–372.

[8] H. Silverman, Neighborhoods of classes of analytic function, *Far East. J. Math. Sci.*, **3**(2) (1995), 165–169.

[9] H.M. Srivastava, S. Owa and S.K. Chatterjea, A note on certain classes of starlike functions, *Rend. Sem. Mat. Univ. Padova*, **77** (1970), 115–124.

[10] H.M. Srivastava and S. Owa, *Current Topics in Analytic Function Theory*, World Scientific, Singapore, (1992).

[11] T. Rosy, K.G. Subramaniyam and G. Murugusundaramoorthy, Neighbourhoods of certain classes of analytic functions with negative coefficients, preprint, presented at 13th– *Annual Conference of the Ramanujan Mathematical Society*, Tirunelveli, TamilNadu, June 1998.

In: Advances in Inequalities for Series
Editors: S. S. Dragomir and A. Sofo, pp. 199–208
ISBN 978-1-60021-920-7

Chapter 14

Generalized Hypergeometric Functions Associated with k-uniformly Convex and k-starlike Functions Defined by Generalized Derivative

T. N. Shanmugam[1], *S. Sivasubramanian*[2] *and M. Darus*[3]
[1]Department of Mathematics, Anna University
Chennai 600025, Tamilnadu, India
E-mail address: drtns2001@yahoo.com
[2]Department of Mathematics, Easwari Engineering College
Chennai-600 089, Tamilnadu, India
E-mail address: sivasaisastha@rediffmail.com
[3]School of Mathematical Sciences, Faculty of Science and Technology
Universiti Kebangsaan Malaysia, Bangi 43600, Selangor D. Ehsan, Malaysia
E-mail address: maslina@pkrisc.cc.ukm.my

Abstract

For a certain linear operator which is defined by means of the Hadamard product with a generalized hypergeometric function, the authors aim at investigating several mapping properties involving certain subclasses of analytic and univalent functions. Relevant connections of the definitions and results presented in this paper with those in several earlier works on the subject are also pointed out.

1991 Mathematics Subject Classification: 30C45

Key words and phrases: Analytic functions, Univalent, Generalized hypergeometric functions, Hadamard product (or convolution), Parabolic starlike function.

1. Introduction

Let $\mathcal{A}$ be the class consisting of functions of the form

$$f(z) = z + \sum_{n=2}^{\infty} a_n z^n, \tag{1.1}$$

which are *analytic and univalent* in the open unit disk $\mathcal{U} := \{z : |z| < 1\}$. We denote by $\mathcal{S}$, the subclass of $\mathcal{A}$ consisting of functions that are also univalent in $\mathcal{U}$.

A function f of the form (1.1) is said to be in the class $k - UCV$, the class of functions that are *k-uniformly convex* in $\mathcal{U}$ if

$$1 + \Re\left\{\frac{zf''(z)}{f'(z)}\right\} \geq k\left|\frac{zf''(z)}{f'(z)}\right| \quad (0 \leq k < \infty). \tag{1.2}$$

A function f of the form (1.1) is said to be in the class $k - ST$, the class of functions that are *k-starlike* in $\mathcal{U}$ if

$$\Re\left\{\frac{zf'(z)}{f(z)}\right\} > k\left|\frac{zf'(z)}{f(z)} - 1\right| \quad (0 \leq k < \infty). \tag{1.3}$$

The class $k - UCV$ was introduced by Kanas and Wiśniowska [8], where its geometrical definition and connections with the conic domains were considered. The class $k - ST$, was investigated by Kanas and Wiśniowska [9] and further properties were studied by Kanas and Srivastava [10]. In particular, when $k = 1$, we obtain

$$1 - UCV \equiv UCV \quad \text{and} \quad 1 - ST \equiv SP,$$

where UCV and SP are the familiar classes of uniformly convex functions and parabolic starlike functions in $\mathcal{U}$, respectively (see, for details, [5, 6, 12] and [16]). In fact, by making use of a fractional calculus operator, Srivastava and Mishra [19] presented a systematic and unified study of the classes UCV and SP.

Ruscheweyh [16] introduced the operator $\mathcal{D}^\mu : \mathcal{A} \to \mathcal{A}$, defined by the Hadamard product (or convolution)

$$\mathcal{D}^\mu f(z) := f(z) * \frac{z}{(1-z)^{\mu+1}} \quad (\mu \geq -1, \quad z \in \mathcal{U}) \tag{1.4}$$

which implies that

$$\mathcal{D}^n f(z) = \frac{z\left(z^{n-1} f(z)\right)^{(n)}}{n!} \quad (n \in \mathcal{N}_0),$$

$$\mathcal{D}^0 f(z) = f(z), \quad \mathcal{D}^1 f(z) = zf'(z), \quad \mathcal{D}^2 f(z) = zf'(z) + \frac{z^2 f''(z)}{2}.$$

We observe that the power series of $\mathcal{D}^\mu f(z)$ for the function f of the form (1.1), in view of (1.4), is given by

$$\mathcal{D}^\mu f(z) = z + \sum_{n=2}^{\infty} \frac{\Gamma(n+\mu)}{(n-1)!\Gamma(1+\mu)} a_n z^n \quad (z \in \mathcal{U}). \tag{1.5}$$

Using the Ruscheweyh derivative, Kanas and Yuguchi [11] introduced the class $UR(\mu,k)$ as

$$UR(\mu,k) := \left\{ f \in \mathcal{A} : \Re\left\{ \frac{z(\mathcal{D}^{\mu} f(z))'}{\mathcal{D}^{\mu} f(z)} \right\} > k \left| \frac{z(\mathcal{D}^{\mu} f(z))'}{\mathcal{D}^{\mu} f(z)} - 1 \right| \quad (z \in \mathcal{U}; 0 \leq k < \infty) \right\}. \tag{1.6}$$

It is easy to check that $UR(1,k) \equiv k-UCV$, and $UR(0,k) \equiv k-ST$.

Next, we define the classes S_{λ}^{*} and C_{λ} (cf., e.g., [14, p. 142]) by

$$S_{\lambda}^{*} := \left\{ f \in S : \left| \frac{zf'(z)}{f(z)} - 1 \right| < \lambda, \quad (z \in \mathcal{U}, \lambda > 0) \right\}$$

and

$$C_{\lambda} := \left\{ f \in S : \left| \frac{zf''(z)}{f'(z)} \right| < \lambda, \quad (z \in \mathcal{U}, \lambda > 0) \right\}.$$

Obviously $f(z) \in C_{\lambda} \Longleftrightarrow zf'(z) \in S_{\lambda}^{*} \quad (\lambda > 0)$ which is analogous to the Alexander equivalence (see, for details, [2]).

Let $\tau \in \mathbb{C} \setminus \{0\}$, $-1 \leq B < A \leq 1$. A function $f \in \mathcal{A}$ is said to be in the class $\mathcal{R}^{\tau}(A,B)$ if it satisfies the inequality

$$\left| \frac{f'(z) - 1}{(A-B)\tau - B[f'(z) - 1]} \right| < 1, \quad (z \in \mathcal{U}). \tag{1.7}$$

The class $\mathcal{R}^{\tau}(A,B)$ was introduced by Dixit and Pal [1]. For the choices of

$$\tau = e^{-i\eta} \cos\eta \left(-\frac{\pi}{2} < \eta < \frac{\pi}{2} \right), \quad A = 1 - 2\gamma \quad (0 \leq \gamma < 1)$$

and $B = -1$, the class $\mathcal{R}^{\tau}(A,B)$ reduces to the class $\mathcal{R}_{\eta}(\gamma)$ studied by Ponnusamy and Rønning [14], where

$$\mathcal{R}_{\eta}(\gamma) := \left\{ f \in \mathcal{A} : \Re\left(e^{i\eta}(f'(z) - \gamma) \right) > 0 \left(z \in \mathcal{U}; -\frac{\pi}{2} < \eta < \frac{\pi}{2}; 0 \leq \gamma < 1 \right) \right\}. \tag{1.8}$$

Finally, we recall a sufficiently adequate special case of a convolution operator which was introduced earlier by Dziok and Srivastava [3] by means of the Hadamard product (or convolution) involving generalized hypergeometric functions.

Let $\alpha_1, \alpha_2, \ldots, \alpha_p$ and $\beta_1, \beta_2, \ldots, \beta_q$ be complex numbers with $\beta_j \neq 0, -1, -2, \ldots; (j = 1, 2, \ldots, q)$. Then the generalized hypergeometric function ${}_pF_q(z)$ is defined by

$$\begin{aligned} {}_pF_q(z) &= {}_pF_1(\alpha_1, \alpha_2, \ldots, \alpha_p; \beta_1, \beta_2, \ldots, \beta_q; z) \\ &= \sum_{n=0}^{\infty} \frac{(\alpha_1)_n (\alpha_2)_n \cdots (\alpha_p)_n z^n}{(\beta_1)_n (\beta_2)_n \cdots (\beta_q)_n (1)_n}, \quad p \leq q+1 \text{ and } \quad p, q \in \mathcal{N}_0 = \mathcal{N} \cup \{0\}, \end{aligned} \tag{1.9}$$

where $(\lambda)_n$ is the *Pochhammer* symbol (or the shifted factorial since $(1)_n = n!$) defined by

$$(\lambda)_n = \begin{cases} 1 & \text{for } n = 0 \\ \lambda(\lambda+1)\cdots(\lambda+n-1) & \text{for } n = 1, 2, 3 \ldots \end{cases} \tag{1.10}$$

We observe that the series in (1.9) converges absolutely in the entire complex plane for $p < q+1$, and in the unit disc for $p = q+1$.

Thus, we have,

$$\left(I \begin{array}{c}\alpha_1,\ldots,\alpha_p\\ \beta_1,\ldots,\beta_q\end{array} f\right)(z) := z\, {}_pF_q(\alpha_1,\ldots,\alpha_p;\beta_1,\ldots,\beta_q,z) * f(z) \tag{1.11}$$

$$(f \in \mathcal{A};\quad p \le q+1;\quad z \in U),$$

so that, for a function of the form (1.1)

$$\left(I \begin{array}{c}\alpha_1,\ldots,\alpha_p\\ \beta_1,\ldots,\beta_q\end{array} f\right)(z) = z + \sum_{n=2}^{\infty} B_n a_n z^n, \tag{1.12}$$

where, for convenience,

$$B_n := \frac{(\alpha_1)_{n-1}\cdots(\alpha_p)_{n-1}}{(\beta_1)_{n-1}\cdots(\beta_q)_{n-1}(1)_{n-1}}, \quad (n \in \mathcal{N}\setminus\{1\}). \tag{1.13}$$

Several mapping properties of the class $k-UCV$ and $k-ST$ using the generalized hypergeometric functions was studied recently by Gangadharan et al. [4]. Just as it was observed by Dziok and Srivastava [3, pp. 3-4], the convolution operator defined by (1.12) includes, as its further special cases, various other linear operators which were considered in many earlier works. In particular, for $p = 2$ and $q = 1$, we obtain the linear operator $\mathcal{F}(\alpha,\beta,\gamma)$ defined by

$$\mathcal{F}(\alpha,\beta,\gamma)f(z) = z\,{}_2F_1(\alpha,\beta,\gamma;z) * f(z) = \left(I_\gamma^{\alpha,\beta} f\right)(z),$$

which was investigated by Hohlov [7]. Also, many univalence, starlikeness, and convexity properties of the hypergeometric functions

$$z\,{}_2F_1(\alpha,\beta,\gamma;z)$$

and

$$z\,{}_pF_q(\alpha_1,\ldots,\alpha_p;\beta_1,\ldots,\beta_q;z) \quad (p \le q+1)$$

were investigated in a number of earlier works (see [13]) (see also [18]).

Our main objective in the present paper is to make use of the linear operator defined in (1.12) in order to establish some connections between the classes $UR(\mu,k)$ and $\mathcal{R}^\tau(A,B)$. Our results generalizes the results obtained by Gangadharan et al. [4].

To prove the main theorems we need the following lemmas.

Lemma 1.1 (see [1]). *If $f \in \mathcal{R}^\tau(A,B)$ is of form (1.1), then*

$$|a_n| \le (A-B)\frac{|\tau|}{n}, \quad n \ne 1. \tag{1.14}$$

Lemma 1.2 (see [11]). *Let $0 \leq k < \infty$, and let $f \in \mathcal{A}$ be of the form (1.1). If*

$$\sum_{n=2}^{\infty} \frac{\Gamma(n+\mu)}{(n-1)!\Gamma(1+\mu)} [n(k+1)-k]\,|a_n| < 1, \tag{1.15}$$

holds true, then $f \in UR(\mu,k)$.

Lemma 1.3 (see [4]). *Let $f \in \mathcal{A}$ be of the form (1.1) . If*

$$\sum_{n=2}^{\infty} (\lambda+n-1)|a_n| \leq \lambda, \quad (\lambda > 0) \tag{1.16}$$

then $f \in S^*_\lambda$.

Lemma 1.4 (see [4]). *Let $f \in \mathcal{A}$ be of the form (1.1). If*

$$\sum_{n=2}^{\infty} n(\lambda+n-1)|a_n| \leq \lambda, \quad (\lambda > 0) \tag{1.17}$$

holds true, then $f \in C_\lambda$.

Lemma 1.5 (see [11]). *Let $0 \leq k < \infty$, and let $f \in \mathcal{A}$ be of the form (1.1). If $f \in UR(\mu,k)$, then*

$$|a_n| \leq \frac{(P_1)_{n-1}\Gamma(1+\mu)}{\Gamma(n+\mu)}, \quad n \in N\backslash\{1\}, \tag{1.18}$$

where $P_1 = P_1(k)$ is the coefficient of z in the function

$$p_k(z) = 1 + \sum_{n=1}^{\infty} P_n(k) z^n, \tag{1.19}$$

which is the extremal function for the class $P(p_k)$ related to the class $UR(\mu,k)$ by the range of the expression

$$1 + \frac{zf''(z)}{f'(z)} (z \in U),$$

where $P_1 = P_1(k)$ is given, as above, by (1.19).

2. Main Results

Theorem 2.1. *Suppose that $\alpha_i \in \mathbb{C} \setminus \{0\} (i = 1,2,3,\ldots,p)$, $\Re(\beta_j) > 0$ $(j = 1,2,3,\ldots,q)$ and (in the case $p = q+1$)*

$$\Re\left(\sum_{j=1}^{q} \beta_j\right) > P_1 - \mu + \sum_{i=1}^{p} |\alpha_i|,$$

where $P_1 = P_1(k)$ is as given by (1.19). If $f \in UR(\mu,k)$, and for some k $(0 \leq k < \infty)$, the hypergeometric inequality

$${}_{p+2}F_{q+2}(|\alpha_1|,\ldots,|\alpha_p|,P_1,\lambda+1;\Re(\beta_1),\ldots,\Re(\beta_q),\lambda,1+\mu;1) < 2, \tag{2.1}$$

holds true, then $\left(I\,{}^{\alpha_1,\ldots,\alpha_p}_{\beta_1,\ldots,\beta_q} f\right) \in S^*_\lambda$.

Proof. In view of Lemma 1.4, it suffices to show, for $f \in UR(\mu,k)$ of the form (1.1), that

$$\sum_{n=2}^{\infty}(\lambda+n-1)|a_n|B_n \le \lambda, \quad (\lambda>0),$$

where B_n is defined by (1.12). By applying the coefficient inequality (1.18) and (1.13) we get,

$$\begin{aligned}
&\sum_{n=2}^{\infty}(\lambda+n-1)|a_n||B_n| \\
&\le \sum_{n=2}^{\infty}(\lambda+n-1)\frac{(P_1)_{n-1}\Gamma(1+\mu)}{\Gamma(n+\mu)}\frac{(|\alpha_1|)_{n-1}\cdots(|\alpha_p|)_{n-1}}{(\Re(\beta_1))_{n-1}\cdots(\Re(\beta_q))_{n-1}(1)_{n-1}} \\
&=\lambda\sum_{n=2}^{\infty}\frac{(1+\lambda)_{n-1}}{(\lambda)_{n-1}}\frac{(P_1)_{n-1}}{(1+\mu)_{n-1}}\frac{(|\alpha_1|)_{n-1}\cdots(|\alpha_p|)_{n-1}}{(\Re(\beta_1))_{n-1}\cdots(\Re(\beta_q))_{n-1}(1)_{n-1}} \\
&=\lambda\left\{{}_{p+2}F_{q+2}(|\alpha_1|,\ldots,|\alpha_p|,P_1,\lambda+1;\Re(\beta_1),\ldots,\Re(\beta_q),1+\mu,\lambda;1)-1\right\} \\
&<\lambda,
\end{aligned}$$

by virtue of hypothesis (2.1). This evidently completes the proof of Theorem 2.1. □

If $\mu=0$, we get the following result obtained by Gangadharan et al.[4, Theorem 8, p. 1525].

Corollary 2.2. *Suppose that* $\alpha_i \in \mathbb{C}\setminus\{0\}(j=1,2,3,\ldots,p)$, $\Re(\beta_j)>0\ (j=1,2,3,\ldots,q)$ *and (in the case* $p=q+1$*)*

$$\Re\left(\sum_{j=1}^{q}\beta_j\right) > P_1+\sum_{i=1}^{p}|\alpha_i|,$$

where $P_1=P_1(k)$ *is as given by (1.19). If* $f\in k-ST$*, and for some* $k\quad(0\le k<\infty)$*, the hypergeometric inequality*

$${}_{p+2}F_{q+2}(|\alpha_1|,\ldots,|\alpha_p|,P_1,\lambda+1;\Re(\beta_1),\ldots,\Re(\beta_q),\lambda,1;1)<2 \quad (\lambda>0), \tag{2.2}$$

holds true, then

$$\left(I\begin{matrix}\alpha_1,\ldots,\alpha_p\\ \beta_1,\ldots,\beta_q\end{matrix} f\right)\in S^*_\lambda.$$

If $\mu=1$ in Theorem 2.1, we get the following result obtained by Gangadharan et al.[4, Theorem 7, p. 1524].

Corollary 2.3. *Suppose that* $\alpha_i \in \mathbb{C}\setminus\{0\}(j=1,2,3,\ldots,p)$, $\Re(\beta_j)>0\ (j=1,2,3,\ldots,q)$ *and (in the case* $p=q+1$*)*

$$\Re\left(\sum_{j=1}^{q}\beta_j\right) > P_1-1+\sum_{i=1}^{p}|\alpha_i|,$$

where $P_1 = P_1(k)$ is as given by (1.19). If $f \in k-UCV$, and for some k $(0 \leq k < \infty)$, the hypergeometric inequality

$$ {}_{p+2}F_{q+2}(|\alpha_1|,\ldots,|\alpha_p|,P_1,\lambda+1;\Re(\beta_1),\ldots,\Re(\beta_q),\lambda,2;1) < 2 \quad (\lambda > 0), \tag{2.3} $$

holds true, then

$$ I\begin{matrix}\alpha_1,\ldots,\alpha_p\\ \beta_1,\ldots,\beta_q\end{matrix} f \in S_\lambda^*. $$

Theorem 2.4. *Suppose that $\alpha_i \in \mathbb{C}\setminus\{0\}$ $(i = 1,2,3,\ldots,p)$, $\Re(\beta_j) > 0$ $(j = 1,2,3,\ldots,q)$ and (in the case $p = q+1$)*

$$ \Re\left(\sum_{j=1}^{q}\beta_j\right) > P_1 - \lambda + \sum_{i=1}^{p}|\alpha_i|, $$

where $P_1 = P_1(k)$ is as given by (1.19). If $f \in UR(\mu,k)$, and for some k $(0 \leq k < \infty)$, the hypergeometric inequality

$$ {}_{p+3}F_{q+3}(|\alpha_1|,\ldots,|\alpha_p|,P_1,\lambda+1,2;\Re(\beta_1),\ldots,\Re(\beta_q),1+\mu,\lambda,1;1) < 2 \tag{2.4} $$

holds true, then

$$ \left(I\begin{matrix}\alpha_1,\ldots,\alpha_p\\ \beta_1,\ldots,\beta_q\end{matrix} f\right) \in C_\lambda. $$

Proof. In view of Lemma 1.1, it suffices to show, for $f \in K-UCV$ of the form (1.1), that

$$ \sum_{n=2}^{\infty} n(\lambda+n-1)|a_n|B_n \leq \lambda, \quad (\lambda > 0), $$

where B_n is defined by (1.6). By applying the coefficient inequality (1.18), we get

$$ \begin{aligned} &\sum_{n=2}^{\infty} n(\lambda+n-1)|a_n||B_n| \\ &\leq \sum_{n=2}^{\infty}(\lambda+n-1)\frac{(P_1)_{n-1}\Gamma(1+\mu)}{\Gamma(n+\mu)}\frac{(|\alpha_1|)_{n-1}\cdots(|\alpha_p|)_{n-1}}{(\Re(\beta_1))_{n-1}\cdots(\Re(\beta_q))_{n-1}(1)_{n-1}} \\ &= \lambda\sum_{n=2}^{\infty} n\frac{(1+\lambda)_{n-1}}{(\lambda)_{n-1}}\frac{(P_1)_{n-1}}{(1+\mu)_{n-1}}\frac{(|\alpha_1|)_{n-1}\cdots(|\alpha_p|)_{n-1}}{(\Re(\beta_1))_{n-1}\cdots(\Re(\beta_q))_{n-1}(1)_{n-1}} \\ &= \lambda\left\{{}_{p+3}F_{q+3}(|\alpha_1|,\ldots,|\alpha_p|,P_1,\lambda+1,2;\Re(\beta_1),\ldots\Re(\beta_q),1+\mu,\lambda,1;1) - 1\right\} \\ &< \lambda. \end{aligned} $$

by virtue of hypothesis (2.4). This evidently completes the proof of Theorem 2.4. □

As we remarked earlier, $\mu = 0$ and $\mu = 1$ in Theorem 2.4, yields the results obtained by Gangadharan et al. [4].

Making use of the coefficient inequality (1.14), in place of the coefficient inequality (1.18), we can establish the following result (Theorem 2.5 below), by appealing to Lemma 1.2.

Theorem 2.5. *Suppose that* $\alpha_i \in \mathbb{C} \setminus \{0\}$ $(i = 1,2,3,\ldots,p)$, $\Re(\beta_j) > 0$ $(j = 1,2,3,\ldots,q)$ *and (in the case* $p = q+1$*)*

$$\Re\left(\sum_{j=1}^{q} \beta_j\right) > \mu + \sum_{i=1}^{p} |\alpha_i|,$$

where $P_1 = P_1(k)$ *is as given by (1.19). If* $f \in \mathcal{R}^{\tau}(A,B)$, *and for some* k $(0 \le k < \infty)$, *the hypergeometric inequality*

$$\begin{aligned}(1+k)&\left\{{}_{p+1}F_{q+1}(|\alpha_1|,\ldots,|\alpha_p|,1+\mu;\Re(\beta_1),\ldots\Re(\beta_q),1;1)\right\}\\ &-k\left\{{}_{p+2}F_{q+2}(|\alpha_1|,\ldots,|\alpha_p|,1+\mu,1;\Re(\beta_1),\ldots\Re(\beta_q),1,2;1)\right\}\\ &< 1+\frac{1}{(A-B)\,|\tau|} \qquad (2.5)\end{aligned}$$

holds true, then

$$\left(I\begin{matrix}\alpha_1,\ldots,\alpha_p\\ \beta_1,\ldots,\beta_q\end{matrix} f\right) \in UR(\mu,k).$$

If $\mu = 0$, we get the corrected version of the result obtained by Gangadharan et al. [4, Theorem 2, p. 1520].

Corollary 2.6. *Suppose that* $\alpha_i \in \mathbb{C}\backslash\{0\}$ $(j = 1,2,3,\ldots,p)$, $\Re(\beta_j) > 0$ $(j = 1,2,3,\ldots,q)$ *and (in the case* $p = q+1$*)*

$$\Re\left(\sum_{j=1}^{q} \beta_j\right) > \sum_{i=1}^{p} |\alpha_i|.$$

If $f \in \mathcal{R}^{\tau}(A,B)$ *and the hypergeometric inequality*

$$\begin{aligned}(1+k)\quad &{}_{p}F_{q}(|\alpha_1|,\ldots,|\alpha_p|;\ \Re(\beta_1),\ldots,\ \Re(\beta_q);1)\\ -k\quad &{}_{p+1}F_{q+1}(|\alpha_1|,\ldots,|\alpha_p|,1;\ \Re(\beta_1),\ldots,\ \Re(\beta_q),2;1) \le \frac{1}{(A-B)|\tau|}+1 \qquad (2.6)\end{aligned}$$

holds true, then

$$\left(I\begin{matrix}\alpha_1,\ldots,\alpha_p\\ \beta_1,\ldots,\beta_q\end{matrix} f\right) \in k-ST.$$

As we remarked earlier, for $\mu = 1$, we get the result obtained by Gangadharan et al. [4].

Bibliography

[1] K.K. Dixit and S.K. Pal, On a class of univalent functions related to complex order, *Indian J. Pure Appl. Math.*, **26**(9) (1995), 889–896.

[2] P.L. Duren, *Univalent Functions*, Volume 259, grundlehren der Mathematischen wissenschaften, Bd., Springer-Verlag, Newyork, (1983).

[3] J. Dziok and H.M. Srivastava, Classes of analytic functions associated with the generalized hypergeometric function, *Appl. Math. Comput.*, **103**(1) (1999), 1–13.

[4] A. Gangadharan, T.N. Shanmugam and H.M. Srivastava, Generalized hypergeometric functions associated with k-uniformly convex functions, *Comput. Math. Appl.*, **44**(12) (2002), 1515–1526.

[5] A.W. Goodman, On uniformly convex functions, *Ann. Polon. Math.*, **56**(1) (1991), 87–92.

[6] A.W. Goodman, On uniformly starlike functions, *J. Math. Anal. Appl.*, **155**(2) (1991), 364–370.

[7] Ju. E. Hohlov, Operators and operations on the class of univalent functions, *Izv. Vyssh. Uchebn. Zaved. Mat.*, no. 10(197) (1978), 83–89.

[8] S. Kanas and A. Wísniowska, Conic regions and k-uniform convexity, *J. Comput. Appl. Math.*, **105**(1-2) (1999), 327–336.

[9] S. Kanas and A. Wiśniowska, Conic domains and starlike functions, *Rev. Roumaine Math. Pures Appl.*, **45**(4) (2000), 647–657.

[10] S. Kanas and H.M. Srivastava, Linear operators associated with k-uniformly convex functions, *Integral Transform. Spec. Funct.*, **9**(2) (2000), 121–132.

[11] S. Kanas and T. Yaguchi, Subclasses of k-uniformly convex and starlike functions defined by generalized derivative. II, *Publ. Inst. Math.* (Beograd) (N.S.) **69(83)** (2001), 91–100.

[12] W.C. Ma and D. Minda, Uniformly convex functions, *Ann. Polon. Math.*, **57**(2) (1992), 165–175.

[13] S. Owa and H.M. Srivastava, Univalent and starlike generalized hypergeometric functions, *Canad. J. Math.*, **39**(5) (1987), 1057–1077.

[14] S. Ponnusamy and F. Rønning, Starlikeness properties for convolutions involving hypergeometric series, *Ann. Univ. Mariae Curie-Skłodowska Sect. A,* **52**(1) (1998), 141–155.

[15] F. Rønning, Uniformly convex functions and a corresponding class of starlike functions, *Proc. Amer. Math. Soc.*, **118**(1) (1993), 189–196.

[16] St. Ruscheweyh, New Criteria for univalent functions, *Proc. Amer. Math. Soc.*, **49** (1975), 109–115.

[17] St. Ruscheweyh and V. Singh, On the order of starlikeness of hypergeometric functions, *J. Math. Anal. Appl.*, **113**(1) (1986), 1–11.

[18] H.M. Srivastava and S. Owa ,*Current Topics in Analytic Function Theory*, World Sci. Publishing, River Edge, NJ, 1992.

[19] H.M. Srivastava and A.K. Mishra, Applications of fractional calculus to parabolic starlike and uniformly convex functions, *Comput. Math. Appl.*, **39**(3-4) (2000), 57–69.

In: Advances in Inequalities for Series ISBN 978-1-60021-920-7
Editors: S. S. Dragomir and A. Sofo, pp. 209–230

Chapter 15

Identities and Estimates Involving Reciprocals of Binomial Coefficients

Anthony Sofo
School of Computer Science and Mathematics
Victoria University, PO Box 14428 Melbourne
VIC 8001, Australia
E-mail address: anthony.sofo@vu.edu.au

Abstract

In this paper we establish some identities involving sums of reciprocals of binomials coefficients. We subsume and extend some results obtained by Sury, Wang and Zhao and give bounds on the resulting series which depend on several parameters.

1991 Mathematics Subject Classification: Primary 11B65

Key words and phrases: Binomial coefficients, combinatorial identities, integral representations.

1. Introduction

The binomial coefficients are defined by

$$\binom{n}{m} = \begin{cases} \dfrac{n!}{m!\,(n-m)!}; & n \geq m, \\ 0; & n < m \end{cases}$$

for n and m positive integers, or, more generally,

$$\binom{z}{w} = \frac{\Gamma(z+1)}{\Gamma(w+1)\,\Gamma(z-w+1)}$$

for z and w non-negative integers, where $\Gamma(x)$ is the Gamma function.

Binomial coefficients play an important role in many areas of mathematics, including number theory, statistics and probability.

Recently, Batir [2], [3] considered series of the form

$$S(a,k)=\sum_{n=1}^{\infty}\frac{1}{n^k\binom{an}{n}}$$

for $a=2$ and 3 and was able to give some closed form expressions through integral representations of $S(a,k)$.

Batir's motivation was aroused through the known Riemann-zeta series

$$\zeta(4)=\frac{36}{17}\sum_{n=1}^{\infty}\frac{1}{n^4\binom{2n}{n}}=\frac{36}{17}S(2,4)=\frac{\pi^4}{90},$$

$$\zeta(3)=\frac{5}{2}\sum_{n=1}^{\infty}\frac{(-1)^{n-1}}{n^3\binom{2n}{n}},\quad \text{and}$$

$$\zeta(2)=\sum_{n=1}^{\infty}\frac{1}{n^2\binom{2n}{n}}=3S(2,2)=\frac{\pi^2}{6}$$

where

$$\zeta(s)=\sum_{n=1}^{\infty}\frac{1}{n^s},\quad \operatorname{Re}(s)>1$$

is the Riemann-zeta function.

Indeed, Apéry [1] used the series representation to prove the irrationality of $\zeta(2)$ and $\zeta(3)$.

Reciprocals of binomial coefficients are also prolific in the mathematical literature. Mansour and West [4] set up the problem: Let B_n be the hyperoctahedral group, the set of all signed permutations on n letters, and let $B_n(T)$ be the set of all signed permutations in B_n which avoid a set T of signed pattern. They then show that some of the cardinalities encountered involve reciprocals of binomial coefficients.

Weinzierl [14] states that in the calculation of higher order corrections to scattering processes in particle physics one encounters higher order transcendental functions. In the expansion of transcendental functions in a small parameter around rational numbers Weinzierl develops an algorithm which allows for the evaluation of particular sums of reciprocals of binomial coefficients. Many results on reciprocals of binomial coefficient identities may be seen in the papers of Mansour [5], Pla [8], Rockett [9], Sury [11], Sury, Wang and Zhao [12], Trif [13], and Zhao and Wang [16].

It is well known that it is difficult to compute the values of combinatorial sums involving reciprocals of binomial coefficients, so any closed form representations, or any bounding of these sums is of great benefit.

Sury [11] used the Beta function

$$B(p,q)=\frac{\Gamma(p)\Gamma(q)}{\Gamma(p+q)}=\int_0^1 t^{p-1}(1-t)^{q-1}\,dt\quad \text{for } p>0,\ q>0$$

to observe that

$$\begin{aligned}\frac{1}{\binom{n}{m}} &= \frac{m!\,(n-m)!}{n!}\\ &= \frac{\Gamma(m+1)\,\Gamma(n-m+1)}{\Gamma(n+1)}\\ &= (n+1)\int_0^1 t^m(1-t)^{n-m}\,dt\end{aligned}$$

Utilising the integral identity for the inverse binomial coefficients, Sury and Trif further showed that

$$\sum_{m=0}^{n}\frac{1}{\binom{n}{m}} = \frac{n+1}{2^n}\sum_{m=0}^{n}\frac{2^m}{m+1} = \frac{n+1}{2^n}\sum_{j\text{ odd}}\frac{1}{j}\binom{n+1}{j}. \tag{1.1}$$

We can also give the alternating version of (1.1) as follows:

Theorem 1.1. *For n a non-negative integer*

$$\sum_{m=0}^{n}\frac{(-1)^m}{\binom{n}{m}} = \begin{cases}\frac{2(n+1)}{n+2}, & \text{for } n \text{ even},\\ 0, & \text{for } n \text{ odd}.\end{cases}$$

Proof. Using the definition of the Gamma function, we have

$$\begin{aligned}\sum_{m=0}^{n}\frac{(-1)^m}{\binom{n}{m}} &= \sum_{m=0}^{n}(-1)^m\frac{(n+1)\Gamma(m+1)\Gamma(n-m+1)}{\Gamma(n+2)}\\ &= \sum_{m=0}^{n}(-1)^m(n+1)\int_0^1 t^m(1-t)^{n-m}\,dt.\end{aligned}$$

Interchanging sum and integral gives us

$$\begin{aligned}\sum_{m=0}^{n}\frac{(-1)^m}{\binom{n}{m}} &= (n+1)\int_0^1(1-t)^n\sum_{m=0}^{n}(-1)^m\left(\frac{t}{1-t}\right)^m dt\\ &= (n+1)\int_0^1\left[(1-t)^{n+1}+(-1)^n t^{n+1}\right]dt\\ &= \begin{cases}\frac{2(n+1)}{n+2}, & \text{for } n \text{ even},\\ 0, & \text{for } n \text{ odd}.\end{cases}\end{aligned}$$

□

Sury, Wang and Zhao [12] proved the following theorem.

Theorem 1.2. *In the ring of $Q[T]$ of rational polynomials, the identity*

$$\sum_{r=m}^{n}\frac{T^r(1-T)^{n-r}}{\binom{n}{r}} = (n+1)\sum_{r=m}^{n}\frac{T^{n+1}(1-T)^{n-r}}{r+1} + (n+1)\sum_{r=0}^{n-m}\frac{T^{n-r}(1-T)^{n-m+1}}{(m+r+1)\binom{m+r}{r}} \tag{1.2}$$

holds for $m \leq n$. An equivalent form is that for $\lambda \neq -1$

$$\sum_{r=m}^{n} \frac{\lambda^r}{\binom{n}{r}} = (n+1) \sum_{r=0}^{n-m} \frac{\lambda^{m+r}}{(\lambda+1)^{r+1}} \sum_{i=0}^{n-m-r} \binom{n-m-r}{i} \frac{(-1)^i}{m+1+i} + (n+1) \frac{\lambda^{n+1}}{(\lambda+1)^{n+2}} \sum_{r=m}^{n} \frac{(\lambda+1)^{r+1}}{r+1}. \quad (1.3)$$

By the use of Theorem 1.2 and noting that for $|x| < 1$,

$$\sum_{r=1}^{\infty} \frac{(2x)^{2r}}{r\binom{2r}{r}} = \frac{2x\arcsin(x)}{\sqrt{1-x^2}} = \int_0^1 \frac{4x^2 t}{1-4x^2 t(1-t)} dt.$$

Sury, Wang and Zhao [12] showed, among other results, that

$$\sum_{n=0}^{\infty} \frac{1}{(n+1)(n+2)\cdots(n+j)} = \frac{1}{(j-1)(j-1)!}, \quad (1.4)$$

$$\sum_{n=0}^{\infty} \frac{1}{(3n+1)(3n+2)(3n+3)} = \frac{\pi\sqrt{3}-3\ln 3}{12}, \quad (1.5)$$

$$\sum_{n=0}^{\infty} \frac{1}{(4n+1)(4n+2)(4n+3)(4n+4)} = \frac{6\ln 2-\pi}{24} \quad (1.6)$$

and

$$\sum_{n=0}^{\infty} \frac{(-1)^n}{\binom{n+j}{n}} = j2^{j-1}\left(\ln 2 - \sum_{r=1}^{j-1} \frac{1}{r}\right) - j\sum_{r=1}^{j-1} (-1)^r \binom{j-1}{r} \frac{2^{j-1-r}}{r}, \quad \text{for } j = 1,2,\ldots \quad (1.7)$$

Identities (1.4) to (1.7) are reciprocal binomial identities of the form

$$S(a,j) = \frac{1}{j!} \sum_{n=0}^{\infty} \frac{1}{\binom{an+j}{an}} \quad (1.8)$$

and

$$T(a,j) = \frac{1}{j!} \sum_{n=0}^{\infty} \frac{(-1)^n}{\binom{an+j}{an}} \quad (1.9)$$

for $j = 1,2,3\ldots,\ a \in \mathbb{R}^+ \setminus \{0\}$.

The identity (1.4)

$$S(1,j) = \frac{1}{(j-1)(j-1)!}$$

and

$$S(2,j) = \frac{2^{j-2}}{(j-1)!} \left[\ln 2 + \sum_{r=1}^{j-2} (-1)^r \binom{j-2}{r} \left(\frac{2^r-1}{r \cdot 2^r}\right)\right] = \frac{1}{j!}\, {}_3F_2 \left[\begin{matrix} \frac{1}{2}, 1, 1, \\ \frac{1+j}{2}, \frac{2+j}{2} \end{matrix} \middle| 1 \right]$$

and others were also previously given by Sofo [10].

In this paper we shall extend the range of identities for $S(a,j)$, $T(a,j)$, give particular closed form representations to $S\left(\frac{1}{b},j\right)$ and $T\left(\frac{1}{b},j\right)$ for $b=2,3,\ldots$.

We shall also give a generalisation to both $S(a,j)$ and $T(a,j)$, and determine bounds on the series.

2. Identity Representations

Transform techniques are used extensively in the analysis of series and in their representation in closed form.

In his work, Wheelon [15], and later Sofo [10] essentially showed that

$$\begin{aligned} S(a,j) &= \frac{1}{j!}\sum_{n=0}^{\infty}\frac{1}{\binom{an+j}{an}} \\ &= \sum_{n=0}^{\infty}\frac{1}{\prod_{k=1}^{j}(an+k)} \\ &= \frac{1}{(j-1)!}\int_{x=0}^{1}\frac{(1-x)^{j-1}}{1-x^a}dx \\ &= \frac{1}{j!}\,{}_{j+1}F_j\left[\left.\begin{matrix} 1,\frac{1}{a},\frac{2}{a},\frac{3}{a},\ldots,\frac{j}{a} \\ 1+\frac{1}{a},1+\frac{2}{a},1+\frac{3}{a},\ldots,1+\frac{j}{a}\end{matrix}\right| 1\right]. \end{aligned}$$

We can use the ideas of Sury, Wang and Zhao [12] by implementing the Beta function to state the following theorem.

Theorem 2.1. *Let* $a\in\mathbb{R}^+\setminus\{0\}$ *and* $j=2,3,4,\ldots$. *Then*

$$S(a,j) = \frac{1}{j!}\sum_{n=0}^{\infty}\frac{1}{\binom{an+j}{an}} = \frac{1}{j!}\sum_{n=0}^{\infty}\frac{1}{\binom{an+j}{j}} \tag{2.1}$$

$$= \sum_{n=0}^{\infty}\frac{1}{\prod_{k=1}^{j}(an+k)} \tag{2.2}$$

$$= \frac{1}{(j-1)!}\int_{x=0}^{1}\frac{(1-x)^{j-1}}{1-x^a}dx \tag{2.3}$$

$$= \frac{1}{j!}\,{}_{j+1}F_j\left[\left.\begin{matrix} 1,\frac{1}{a},\frac{2}{a},\frac{3}{a},\ldots,\frac{j}{a} \\ 1+\frac{1}{a},1+\frac{2}{a},1+\frac{3}{a},\ldots,1+\frac{j}{a}\end{matrix}\right| 1\right] \tag{2.4}$$

and similarly

$$T(a,j) = \frac{1}{j!}\sum_{n=0}^{\infty}\frac{(-1)^n}{\binom{an+j}{an}} \tag{2.5}$$

$$= \sum_{n=0}^{\infty}\frac{(-1)^n}{\prod_{k=1}^{j}(an+k)} \tag{2.6}$$

$$= \frac{1}{(j-1)!}\int_{x=0}^{1}\frac{(1-x)^{j-1}}{1+x^a}dx \tag{2.7}$$

$$= \frac{1}{j!}\ {}_{j+1}F_j\left[\left.\begin{array}{c} 1,\frac{1}{a},\frac{2}{a},\frac{3}{a},\ldots,\frac{j}{a} \\ 1+\frac{1}{a},1+\frac{2}{a},1+\frac{3}{a},\ldots,1+\frac{j}{a}\end{array}\right| -1\right]. \tag{2.8}$$

Proof. Consider the alternating case

$$\begin{aligned} T(a,j) &:= \frac{1}{j!}\sum_{n=0}^{\infty}\frac{(-1)^n}{\binom{an+j}{an}} = \frac{1}{j!}\sum_{n=0}^{\infty}\frac{(-1)^n}{\binom{an+j}{j}} \\ &= \frac{1}{(j-1)!}\sum_{n=0}^{\infty}(-1)^n\frac{\Gamma(j)\Gamma(an+1)}{\Gamma(an+j+1)} \\ &= \frac{1}{(j-1)!}\sum_{n=0}^{\infty}(-1)^n B(an+1,j) \\ &= \frac{1}{(j-1)!}\sum_{n=0}^{\infty}\int_{x=0}^{1}(-1)^n x^{an}(1-x)^{j-1}\,dx. \end{aligned}$$

Interchanging the sum and integral, we have

$$\begin{aligned} T(a,j) &= \frac{1}{(j-1)!}\int_{x=0}^{1}\sum_{n=0}^{\infty}(-1)^n x^{an}(1-x)^{j-1}\,dx \\ &= \frac{1}{(j-1)!}\int_{x=0}^{1}\frac{(1-x)^{j-1}}{1+x^a}dx \end{aligned}$$

which is the result (2.7). $B(\alpha,\beta)$ is the classical Beta function.

From the ratio of successive terms of (2.5)

$$\frac{V_{n+1}}{V_n} = -\frac{\prod_{j=1}^{k}(an+k)}{\prod_{j=1}^{k}(an+a+k)}$$

we arrive at the result (2.8).

The proof of $S(a,j)$ follows a similar argument as that used for $T(a,j)$ and will not be done here. □

In the next theorem we consider a particular case for the value of a.

Theorem 2.2. *In the case that $a = \frac{1}{b}$, for b an even number, then $T\left(\frac{1}{b}, j\right)$ forms the rational numbers*

$$T\left(\frac{1}{b}, j\right) = \sum_{n=0}^{\infty} \frac{(-1)^n}{\prod_{k=1}^{j}\left(\frac{n}{b}+k\right)} = \frac{1}{(j-1)} \sum_{\mu=0}^{b-1} \frac{(-1)^{\mu}}{\prod_{\nu=1}^{j-1}\left(\nu+\frac{\mu}{b}\right)} \tag{2.9}$$

and similarly for $b = 2,3,4,5\ldots..$, then

$$S\left(\frac{1}{b}, j\right) = \sum_{n=0}^{\infty} \frac{1}{\prod_{k=1}^{j}\left(\frac{n}{b}+k\right)} = \frac{1}{(j-1)} \sum_{\mu=0}^{b-1} \frac{1}{\prod_{\nu=1}^{j-1}\left(\nu+\frac{\mu}{b}\right)}. \tag{2.10}$$

Proof. Consider

$$\begin{aligned}
T\left(\frac{1}{b}, j\right) &= \sum_{n=0}^{\infty} \frac{(-1)^n}{\prod_{k=1}^{j}\left(\frac{n}{b}+k\right)} = \frac{1}{(j-1)!}\int_{x=0}^{1} \frac{(1-x)^{j-1}}{1+x^{\frac{1}{b}}}dx \\
&= \frac{1}{(j-1)!}\int_{x=0}^{1} (1-x)^{j-2}\left(\frac{1-x}{1+x^{\frac{1}{b}}}\right)dx \\
&= \frac{1}{(j-1)!}\int_{x=0}^{1} (1-x)^{j-2}\sum_{\mu=0}^{b-1}(-1)^{\mu}x^{\frac{\mu}{b}}dx \\
&= \frac{1}{(j-1)!}\int_{x=0}^{1} \sum_{r=0}^{j-2}(-1)^r\binom{j-2}{r}x^r\sum_{\mu=0}^{b-1}(-1)^{\mu}x^{\frac{\mu}{b}}dx \\
&= \frac{1}{(j-1)!}\sum_{r=0}^{j-2}(-1)^r\binom{j-2}{r}\sum_{\mu=0}^{b-1}\frac{(-1)^{\mu}}{r+1+\frac{\mu}{b}} \\
&= \frac{1}{(j-1)!}\sum_{\mu=0}^{b-1}\frac{(-1)^{\mu}}{\left(1+\frac{\mu}{b}\right)\binom{j-1+\frac{\mu}{b}}{j-2}} \\
&= \frac{1}{(j-1)}\sum_{\mu=0}^{b-1}\frac{(-1)^{\mu}}{\prod_{\nu=1}^{j-1}\left(\nu+\frac{\mu}{b}\right)},
\end{aligned}$$

which is the result (2.9).

From (2.8) it is also interesting to note that, for b even

$$ {}_{j+1}F_j\left[\begin{matrix} 1, b, 2\cdot b, 3\cdot b, \ldots, j\cdot b \\ 1+b, 1+2\cdot b, 1+3\cdot b, \ldots, 1+j\cdot b \end{matrix}\middle| -1\right] = j\sum_{\mu=0}^{b-1}\frac{(-1)^{\mu}}{\prod_{\nu=1}^{j-1}\left(\nu+\frac{\mu}{b}\right)}$$

The result (2.10) can be proved in the same way and will not be detailed here. Again from (2.4) it is interesting to note that, for b a positive integer

$$ {}_{j+1}F_j\left[\begin{matrix} 1, b, 2\cdot b, 3\cdot b, \ldots, j\cdot b \\ 1+b, 1+2\cdot b, 1+3\cdot b, \ldots, 1+j\cdot b \end{matrix}\middle| 1\right] = j\sum_{\mu=0}^{b-1}\frac{1}{\prod_{\nu=1}^{j-1}\left(\nu+\frac{\mu}{b}\right)}.$$

□

3. Examples

In the case when a is a positive integer, by known properties of the hypergeometric function we may state that

$$ {}_{j+1}F_j\left[\begin{array}{c} 1,\frac{1}{a},\frac{2}{a},\frac{3}{a},\ldots,\frac{j}{a} \\ 1+\frac{1}{a},1+\frac{2}{a},1+\frac{3}{a},\ldots,1+\frac{j}{a} \end{array}\middle| -1\right] = {}_{a+1}F_a\left[\begin{array}{c} 1,\frac{1}{a},\frac{2}{a},\frac{3}{a},\ldots,\frac{a}{a} \\ \frac{1+j}{a},\frac{2+j}{a},\frac{3+j}{a},\ldots,\frac{a+j}{a} \end{array}\middle| -1\right]. $$

Consequently, we may write

$$ T(a,j) = \sum_{n=0}^{\infty} \frac{(-1)^n}{\prod_{k=1}^{j}(an+k)} = \frac{1}{j!}\ {}_{a+1}F_a\left[\begin{array}{c} 1,\frac{1}{a},\frac{2}{a},\frac{3}{a},\ldots,\frac{a}{a} \\ \frac{1+j}{a},\frac{2+j}{a},\frac{3+j}{a},\ldots,\frac{a+j}{a} \end{array}\middle| -1\right]. $$

(i) For $a=1$ we have,

$$ \begin{aligned} T(1,j) &= \sum_{n=0}^{\infty} \frac{(-1)^n}{\prod_{k=1}^{j}(n+k)} \\ &= \frac{1}{j!}\ {}_2F_1\left[\begin{array}{c} 1,1 \\ 1+j \end{array}\middle| -1\right] \\ &= \frac{2^{j-1}\ln 2}{(j-1)!} + \frac{2^{j-1}}{(j-1)!}\sum_{r=1}^{j-1}(-1)^r\binom{j-1}{r}\left(\frac{2^r-1}{r\cdot 2^r}\right) \\ &= \begin{cases} 2\ln 2 - 1; & \text{for } j=2 \\ \frac{2^{j-2}}{(j-2)!}\left\{ {}_3F_2\left[\begin{array}{c} 1,1,2-j \\ 2,2 \end{array}\middle| -1\right] - {}_3F_2\left[\begin{array}{c} 1,1,2-j \\ 2,2 \end{array}\middle| -\frac{1}{2}\right]\right\}; & \text{for } j>2. \end{cases} \end{aligned} $$

This confirms the result (1.7), of the paper of Sury, Wang and Zhao [12].

(ii) For $a=2$ and $j=4m+5$, $m=0,1,2,\ldots$, we may extract the following result

$$ \begin{aligned} T(2,4m+5) &:= \sum_{n=0}^{\infty} \frac{(-1)^n}{\prod_{k=1}^{4m+5}(2n+k)} \\ &= \frac{1}{(4m+5)!}\ {}_3F_2\left[\begin{array}{c} \frac{1}{2},1,1 \\ 2m+3,\frac{4m+7}{2} \end{array}\middle| -1\right] \\ &= \frac{2^{2m+2}}{(4m+4)!}\left\{(-1)^{m+1}\frac{\pi}{4} + \sum_{p=0}^{m}\frac{(-1)^p}{2m-2p+1}\right. \\ &\quad \left. + \sum_{r=1}^{2m+2}\left(-\frac{1}{2}\right)^r\binom{2m+2}{r}\sum_{s=0}^{r-1}\binom{r-1}{s}\frac{1}{2m+2s+3-r}\right\} \\ &= \frac{-(-1)^m 4^m\pi}{(4m+4)!} + \frac{2^{2m+2}}{(4m+4)!}\left\{\sum_{p=0}^{m}\frac{(-1)^p}{2m-2p+1}\right. \\ &\quad \left. + \sum_{r=1}^{2m+2}\frac{\left(-\frac{1}{2}\right)^r}{2m+3-r}\binom{2m+2}{r}\ {}_2F_1\left[\begin{array}{c} 1-r,m+\frac{3}{2}-\frac{r}{2} \\ m+\frac{5}{2}-\frac{r}{2} \end{array}\middle| -1\right]\right\}. \end{aligned} $$

A rearrangement of the above result highlights an identity for π;

$$(-1)^m \frac{\pi}{4} = \sum_{p=0}^{m} \frac{(-1)^p}{2m-2p+1} + \sum_{r=1}^{2m+2} \left(-\frac{1}{2}\right)^r \binom{2m+2}{r} \sum_{s=0}^{r-1} \frac{\binom{r-1}{s}}{2m+2s+3-r}$$
$$- \frac{(4m+4)!}{2^{2m+2}} \sum_{n=0}^{\infty} \frac{(-1)^n}{\prod_{k=1}^{4m+5}(2n+k)}; \quad m = 0, 1, 2, \ldots.$$

In particular, for $m = 1$

$$\frac{\pi}{4} = \frac{109}{140} + 2520 \sum_{n=0}^{\infty} \frac{(-1)^n}{\prod_{k=1}^{9}(2n+k)}.$$

Note: For $n = 0$, the first term bounds π as follows:

$$\frac{3958}{1260} < \pi < \frac{3959}{1260} < \frac{22}{7}.$$

Remark 3.1. *The series (2.2), $S(a, j)$ and (2.6), $T(a, j)$ can be expressed in terms of the Lerch transcendent.*

In particular, from (2.6)

$$T(a, j) := \sum_{n=0}^{\infty} \frac{(-1)^n}{\prod_{k=1}^{j}(an+k)} = \sum_{n=0}^{\infty} (-1)^n \sum_{k=1}^{j} \frac{A_{j,k}}{(an+k)},$$

where

$$A_{j,k} = \lim_{n \to \left(-\frac{k}{a}\right)} \left\{ \frac{an+k}{\prod_{k=1}^{j}(an+k)} \right\} = \frac{(-1)^{k+1}}{(k-1)!(j-k)!}.$$

Hence

$$T(a, j) = \sum_{n=0}^{\infty} (-1)^n \sum_{k=1}^{j} \frac{(-1)^{k+1}}{(k-1)!(j-k)!(an+k)} \tag{3.1}$$
$$= \sum_{k=1}^{j} \frac{(-1)^{k+1}}{(k-1)!(j-k)!} \sum_{n=0}^{\infty} \frac{(-1)^n}{(an+k)}$$
$$= \frac{1}{2a} \sum_{k=1}^{j} \frac{(-1)^{k+1}}{(k-1)!(j-k)!} \left\{ \psi\left(\frac{1}{2} + \frac{k}{2a}\right) - \psi\left(\frac{k}{2a}\right) \right\},$$

where $\psi(z)$ is the Psi, or digamma function.

The Lerch transcendent, $\phi(z, s, \alpha)$ is defined as

$$\phi(z, s, \alpha) = \sum_{n=0}^{\infty} \frac{z^n}{(n+\alpha)^s},$$

where the $(n+\alpha) = 0$ term is excluded from the sum.

The polygamma functions $\psi^{(k)}(z)$, $k \in \mathbb{N}$ are defined by

$$\psi^{(k)}(z) := \frac{d^{k+1}}{dz^{k+1}} \log \Gamma(z) = \frac{d^k}{dz^k}\left(\frac{\Gamma'(z)}{\Gamma(z)}\right), \quad k \in \mathbb{N}_0 := \mathbb{N} \cup \{0\},$$

where $\psi^{(0)}(z) = \psi(z)$, denotes the Psi, or digamma function, defined by

$$\psi(z) = \frac{d}{dz} \log \Gamma(z) = \frac{\Gamma'(z)}{\Gamma(z)} \quad \text{or} \quad \log \Gamma(z) = \int_1^z \psi(t)\, dt.$$

From (3.1)

$$T(a,j) = \sum_{k=1}^{j} \frac{(-1)^{k+1}}{(k-1)!(j-k)!} \cdot \frac{1}{a} \sum_{n=0}^{\infty} \frac{(-1)^n}{\left(n+\frac{k}{a}\right)} = \sum_{k=1}^{j} \frac{(-1)^{k+1}}{(k-1)!(j-k)!a} \phi\left(-1,1,\frac{k}{a}\right).$$

Similar technical details can be written about the series (2.2), however, they will not be detailed here.

4. Generalisation

Both series (2.1) and (2.5) for $S(a,j)$ and $T(a,j)$ can be generalised in the following manner.

Theorem 4.1. *For $m \geq 1$ and $a > 0$ and j a positive integer, then*

$$S(a,j,m) = \frac{1}{j!} \sum_{n=0}^{\infty} \frac{\binom{n+m-1}{n}}{\binom{an+j}{an}} = \frac{1}{j!} \sum_{n=0}^{\infty} \frac{\binom{n+m-1}{m-1}}{\binom{an+j}{j}} \tag{4.1}$$

$$= \frac{1}{(j-1)!} \int_0^1 \frac{(1-x)^{j-1}}{(1-x^a)^m} dx \tag{4.2}$$

$$= \frac{1}{j!}\, {}_{j+1}F_j \left[\left. \begin{matrix} m, \frac{1}{a}, \frac{2}{a}, \frac{3}{a}, \ldots, \frac{j}{a} \\ 1+\frac{1}{a}, 1+\frac{2}{a}, 1+\frac{3}{a}, \ldots, 1+\frac{j}{a} \end{matrix} \right| 1 \right] \tag{4.3}$$

$$= \sum_{n=0}^{\infty} (m)_n \Big/ n! \prod_{k=1}^{j} (an+k) \tag{4.4}$$

and

$$T(a,j,m) = \frac{1}{j!} \sum_{n=0}^{\infty} \frac{(-1)^n \binom{n+m-1}{n}}{\binom{an+j}{an}} = \frac{1}{j!} \sum_{n=0}^{\infty} \frac{(-1)^n \binom{n+m-1}{m-1}}{\binom{an+j}{j}} \tag{4.5}$$

$$= \frac{1}{(j-1)!} \int_0^1 \frac{(1-x)^{j-1}}{(1+x^a)^m} dx \tag{4.6}$$

$$= \frac{1}{j!}\, {}_{j+1}F_j \left[\left. \begin{matrix} m, \frac{1}{a}, \frac{2}{a}, \frac{3}{a}, \ldots, \frac{j}{a} \\ 1+\frac{1}{a}, 1+\frac{2}{a}, 1+\frac{3}{a}, \ldots, 1+\frac{j}{a} \end{matrix} \right| -1 \right] \tag{4.7}$$

$$= \sum_{n=0}^{\infty} (-1)^n (m)_n \Big/ n! \prod_{k=1}^{j} (an+k). \tag{4.8}$$

Proof. Consider

$$T(a,j,m) = \frac{1}{j!}\sum_{n=0}^{\infty}(-1)^n \frac{\binom{n+m-1}{n}}{\binom{an+j}{an}}$$
$$= \frac{1}{(j-1)!}\sum_{n=0}^{\infty}(-1)^n \binom{n+m-1}{n}\frac{\Gamma(an+1)\Gamma(j)}{\Gamma(an+1+j)}$$
$$= \frac{1}{(j-1)!}\sum_{n=0}^{\infty}(-1)^n \binom{n+m-1}{n} B(an+1,j)$$
$$= \frac{1}{(j-1)!}\sum_{n=0}^{\infty}(-1)^n \binom{n+m-1}{n}\int_0^1 x^{an}(1-x)^{j-1}\,dx,$$

interchanging sum and integral, we have

$$T(a,j,m) = \frac{1}{(j-1)!}\int_0^1 (1-x)^{j-1}\sum_{n=0}^{\infty}(-1)^n \binom{n+m-1}{n} x^{an}dx$$
$$= \frac{1}{(j-1)!}\int_0^1 \frac{(1-x)^{j-1}}{(1+x^a)^m}dx,$$

which is (4.6).

Next, by considering the ratio of terms in (4.5), the hypergeometric identity (4.7) follows. Notice that in the case when a is a positive integer, because of the properties of the hypergeometric function, we may also write

$$T(a,j,m) = \frac{1}{j!}\,{}_{a+1}F_a\left[\begin{matrix} m, \frac{1}{a}, \frac{2}{a}, \frac{3}{a}, \ldots, \frac{a}{a} \\ \frac{1+j}{a}, \frac{2+j}{a}, \frac{3+j}{a}, \ldots, \frac{a+j}{a} \end{matrix}\middle| -1\right].$$

To deduce (4.8) we may write

$$T(a,j,m) = \frac{1}{j!}\sum_{n=0}^{\infty}(-1)^n \frac{\Gamma(m+n)}{\Gamma(m)\Gamma(n+1)}\cdot\frac{j!\Gamma(an+1)}{\Gamma(an+1+j)} = \sum_{n=0}^{\infty}\frac{(-1)^n (m)_n}{n!\,(an+1)_j},$$

where

$$(p)_\alpha = \begin{cases} p(p+1)\cdots(p+\alpha-1) = \frac{\Gamma(p+\alpha)}{\Gamma(p)}, & \text{for } \alpha = 1,2,3,\ldots \\ 1, & \text{for } \alpha = 0 \end{cases}$$

is Pochhammer's symbol.

Now we can state

$$T(a,j,m) = \sum_{n=0}^{\infty}\frac{(-1)^n}{n!}\cdot\frac{(m)_n}{\prod_{k=1}^{j}(an+k)}.$$

The identities (4.2) and (4.3) of $S(a,j,m)$ in (4.1) follow in a similar fashion as above and will not be detailed here. □

Some examples now follow, with the minimum of detail.

Example 3.

$$S(1,j,m)=\frac{1}{j!}\sum_{n=0}^{\infty}\frac{\binom{n+m-1}{n}}{\binom{n+j}{n}}=\frac{1}{(j-m)(j-1)!},$$

hence

$$\sum_{n=0}^{\infty}\frac{\binom{n+m-1}{n}}{\binom{n+j}{n}}=\frac{j}{j-m}\quad for\ \ j\neq m$$

$$={}_2F_1\left[\begin{array}{c}1,m\\1+j\end{array}\middle|1\right],$$

this generalises the result (1.4).

Example 4.

$$S(2,j,m)=\frac{1}{j!}\sum_{n=0}^{\infty}\frac{\binom{n+m-1}{n}}{\binom{2n+j}{2n}},$$

hence

$$\sum_{n=0}^{\infty}\frac{\binom{n+m-1}{n}}{\binom{2n+j}{2n}}=j\cdot 2^{j-1-m}\left[(-1)^{m+1}2^{1-m}\binom{j-1-m}{m-1}\ln 2\right.$$

$$\left.+\sum_{\substack{r=0\\r\neq m-1}}^{j-1-m}(-1)^r\binom{j-1-m}{r}\left(\frac{2^{r+1-m}-1}{2^r(r+1-m)}\right)\right]$$

$$={}_3F_2\left[\begin{array}{c}\frac{1}{2},1,m\\\frac{1+j}{2},\frac{2+j}{2}\end{array}\middle|1\right].$$

Example 5.

$$T(1,j,m)=\frac{1}{j!}\sum_{n=0}^{\infty}(-1)^n\frac{\binom{n+m-1}{n}}{\binom{n+j}{n}},$$

hence

$$\sum_{n=0}^{\infty}(-1)^n\frac{\binom{n+m-1}{n}}{\binom{n+j}{n}}=j\cdot 2^{j-1}\left[(-1)^{m+1}2^{1-m}\binom{j-1}{m-1}\ln 2\right.$$

$$\left.+\sum_{\substack{r=0\\r\neq m-1}}^{j-1-m}(-1)^r\binom{j-1}{r}\left(\frac{2^{r+1-m}-1}{2^r(r+1-m)}\right)\right]$$

$$={}_2F_1\left[\begin{array}{c}1,m\\1+j\end{array}\middle|-1\right].$$

Example 6.

$$S\left(2,9,\frac{9}{2}\right)=\frac{1}{9!}\sum_{n=0}^{\infty}\frac{\binom{n+\frac{7}{2}}{n}}{\binom{2n+9}{2n}},$$

hence

$$\sum_{n=0}^{\infty}\frac{\binom{n+\frac{7}{2}}{n}}{\binom{2n+9}{2n}}=\frac{9\pi}{2}-\frac{456}{35}$$

$$=\sum_{n=0}^{\infty}\frac{\left(\frac{9}{2}\right)_n}{n!\binom{2n+9}{2n}}={}_3F_2\left[\begin{matrix}\frac{1}{2},1,\frac{9}{2}\\ \frac{11}{2},2\end{matrix}\,\middle|\,1\right].$$

Also,

$$T\left(2,9,\frac{9}{2}\right)=\frac{1}{9!}\sum_{n=0}^{\infty}(-1)^n\frac{\binom{n+\frac{7}{2}}{n}}{\binom{2n+9}{2n}},$$

hence

$$\sum_{n=0}^{\infty}\frac{(-1)^n\binom{n+\frac{7}{2}}{n}}{\binom{2n+9}{2n}}=9\ln\left(\sqrt{2}+1\right)+\frac{2559\sqrt{2}}{35}-\frac{552}{5}$$

$$=\sum_{n=0}^{\infty}\frac{(-1)^n\left(\frac{9}{2}\right)_n}{n!\binom{2n+9}{2n}}$$

$$={}_3F_2\left[\begin{matrix}\frac{1}{2},1,\frac{9}{2}\\ \frac{11}{2},2\end{matrix}\,\middle|\,-1\right].$$

Two other instructive examples now follow.

Theorem 4.2. *For m a positive integer, then*

$$S(3,m+1,m)=\sum_{n=0}^{\infty}\frac{\binom{n+m-1}{n}}{\binom{3n+m+1}{3n}}$$

$$=\frac{2\sqrt{3}(m+1)}{3^m}\binom{2m-2}{m-1}\left[\frac{\pi}{6}+\frac{\sqrt{3}}{4}\sum_{r=0}^{m-2}\frac{1-3^r}{(r+1)\binom{2r+1}{r}}\right]$$

$$={}_4F_3\left[\begin{matrix}m,\frac{1}{3},\frac{2}{3},1\\ \frac{m+2}{3},\frac{m+3}{3},\frac{m+4}{3}\end{matrix}\,\middle|\,1\right].$$

Proof.

$$
\begin{aligned}
S(3,m+1,m) &= \frac{1}{m!}\int_0^1 \left(\frac{1-x}{1-x^3}\right)^m dx \\
&= \left(\frac{2}{\sqrt{3}}\right)^{2m-1} \frac{1}{m!}\int_{\theta=\pi/6}^{\pi/3} \cos^{2m-2}\theta d\theta \\
&= \frac{2\sqrt{3}}{m!3^m}\binom{2m-2}{m-1}\left[\theta+\sin\theta\sum_{r=0}^{m-2}\frac{2^{2r}\cos^{2r+1}\theta}{(r+1)\binom{2r+1}{r}}\right]_{\theta=\pi/6}^{\pi/3} \\
&= \frac{2\sqrt{3}}{m!3^m}\binom{2m-2}{m-1}\left[\frac{\pi}{6}+\frac{\sqrt{3}}{4}\sum_{r=0}^{m-2}\frac{1-3^r}{(r+1)\binom{2r+1}{r}}\right]
\end{aligned}
$$

hence

$$
\sum_{n=0}^{\infty}\frac{\binom{n+m-1}{n}}{\binom{3n+m+1}{3n}} = \frac{(m+1)2\sqrt{3}}{3^m}\binom{2m-2}{m-1}\left[\frac{\pi}{6}+\frac{\sqrt{3}}{4}\sum_{r=0}^{m-2}\frac{1-3^r}{(r+1)\binom{2r+1}{r}}\right].
$$

By the properties of hypergeometric functions described earlier, we have

$$
\sum_{n=0}^{\infty}\frac{\binom{n+m-1}{n}}{\binom{3n+m+1}{3n}} = {}_4F_3\left[\begin{matrix} m,\frac{1}{3},\frac{2}{3},1 \\ \frac{m+2}{3},\frac{m+3}{3},\frac{m+4}{3}\end{matrix}\middle| 1\right]
$$

and hence

$$
{}_4F_3\left[\begin{matrix} m,\frac{1}{3},\frac{2}{3},1 \\ \frac{m+2}{3},\frac{m+3}{3},\frac{m+4}{3}\end{matrix}\middle| 1\right] = \frac{2\sqrt{3}}{3^m}\binom{2m-2}{m-1}\left[\frac{\pi}{6}+\frac{\sqrt{3}}{4}\sum_{r=0}^{m-2}\frac{1-3^r}{(r+1)\binom{2r+1}{r}}\right].
$$

Again, by properties of hypergeometric functions

$$
\sum_{n=0}^{\infty}\frac{\binom{n+m-1}{n}}{\binom{3n+m+1}{3n}} = \frac{1}{2(2m-1)m!}\left\{{}_2F_1\left[\begin{matrix}1,m \\ m+\frac{1}{2}\end{matrix}\middle|\frac{3}{4}\right] - 3^{1-m}\,{}_2F_1\left[\begin{matrix}1,m \\ m+\frac{1}{2}\end{matrix}\middle|\frac{1}{4}\right]\right\}.
$$

□

Theorem 4.3. *For* $a=\frac{1}{b}$, $b=2,3,4,\ldots;\ j-1>m$, $m=1,2,3,\ldots$

$$
S\left(\frac{1}{b},j,m\right) = \frac{1}{j!}\sum_{n=0}^{\infty}\frac{\binom{n+m-1}{n}}{\binom{n/b+j}{n/b}}
$$

forms the rational numbers

$$
S\left(\frac{1}{b},j,m\right) = \frac{1}{m!\binom{j-1}{j-1-m}}\sum_{\mu=0}^{m(b-1)}\frac{\alpha_{\mu}^{b,m}}{\prod_{\nu=1}^{j-m}\left(\nu+\frac{\mu}{b}\right)},
$$

where $\alpha_\mu^{b,m}$ are the coefficients of the expansion of the sums of the powers of $x^{\mu/b}$ in $\left(\sum_{\mu=0}^{b-1} x^{\mu/b}\right)^m$.

Similarly for b an even positive integer

$$T\left(\frac{1}{b}, j, m\right) = \frac{1}{j!}\sum_{n=0}^{\infty} \frac{(-1)^n \binom{n+m-1}{n}}{\binom{n/b+j}{n/b}}$$

forms the rational numbers

$$T\left(\frac{1}{b}, j, m\right) = \frac{1}{m!\binom{j-1}{j-1-m}} \sum_{\mu=0}^{m(b-1)} \frac{(-1)^\mu \alpha_\mu^{b,m}}{\prod_{\nu=1}^{j-m}\left(\nu+\frac{\mu}{b}\right)}.$$

Proof.

$$\begin{aligned} S\left(\frac{1}{b}, j, m\right) &= \frac{1}{j!}\sum_{n=0}^{\infty} \frac{\binom{n+m-1}{n}}{\binom{n/b+j}{n/b}} \\ &= \frac{1}{(j-1)!}\int_0^1 \frac{(1-x)^{j-1}}{(1-x^{1/b})^m}dx \\ &= \frac{1}{(j-1)!}\int_0^1 (1-x)^{j-1-m}\left(\frac{1-x}{1-x^{1/b}}\right)^m dx. \end{aligned}$$

Now note that

$$(1-x) = \left(1-x^{1/b}\right)\sum_{\mu=0}^{b-1} x^{\mu/b},$$

so that

$$\begin{aligned} S\left(\frac{1}{b}, j, m\right) &= \frac{1}{(j-1)!}\int_0^1 (1-x)^{j-1-m}\left(\sum_{\mu=0}^{b-1} x^{\mu/b}\right)^m dx \\ &= \frac{1}{(j-1)!}\int_0^1 \sum_{r=0}^{j-1-m} (-1)^r \binom{j-1-m}{r} x^r \sum_{\mu=0}^{m(b-1)} \alpha_\mu^{b,m} x^{\mu/b} dx, \end{aligned}$$

where $\alpha_\mu^{b,m}$ are the coefficients of the expansion of the sum of the powers of $x^{\mu/b}$ in

$\left(\sum_{\mu=0}^{b-1} x^{\mu/b}\right)^m$.

$$\begin{aligned}
S\left(\frac{1}{b}, j, m\right) &= \frac{1}{(j-1)!}\int_0^1 \sum_{\mu=0}^{m(b-1)} \alpha_\mu^{b,m} \sum_{r=0}^{j-1-m} (-1)^r \binom{j-1-m}{r} x^{r+\mu/b} dx \\
&= \frac{1}{(j-1)!} \sum_{\mu=0}^{m(b-1)} \alpha_\mu^{b,m} \sum_{r=0}^{j-1-m} (-1)^r \binom{j-1-m}{r} \frac{1}{r+1+\frac{\mu}{b}} \\
&= \frac{1}{(j-1)!} \sum_{\mu=0}^{m(b-1)} \frac{\alpha_\mu^{b,m}}{\left(1+\frac{\mu}{b}\right)\binom{j-m+\mu/b}{j-1-m}} \\
&= \frac{1}{(j-1)!} \sum_{\mu=0}^{m(b-1)} \frac{\alpha_\mu^{b,m}\Gamma(j-m)\Gamma\left(2+\frac{\mu}{b}\right)}{\left(1+\frac{\mu}{b}\right)\Gamma\left(j-m+1+\frac{\mu}{b}\right)} \\
&= \frac{1}{(j-1)!} \sum_{\mu=0}^{m(b-1)} \frac{\alpha_\mu^{b,m}(j-1-m)!}{\left(1+\frac{\mu}{b}\right)\prod_{\nu=2}^{j-m}(\nu+\mu/b)} \\
&= \frac{1}{m!\binom{j-1}{j-1-m}} \sum_{\mu=0}^{m(b-1)} \frac{\alpha_\mu^{b,m}}{\prod_{\nu=1}^{j-m}\left(\nu+\frac{\mu}{b}\right)}.
\end{aligned}$$

The proof of the result

$$\sum_{n=0}^{\infty} \frac{(-1)^n \binom{n+m-1}{n}}{\binom{n/b+j}{n/b}} = j(j-1-m)! \sum_{\mu=0}^{m(b-1)} \frac{(-1)^\mu \alpha_\mu^{b,m}}{\prod_{\nu=1}^{j-m}\left(\nu+\frac{\mu}{b}\right)}$$

is done in the same fashion as above and will not be detailed here. □

Note:

(1) For $m=1$

$$S\left(\frac{1}{b}, j, 1\right) = \frac{1}{j!}\sum_{n=0}^{\infty} \frac{1}{\binom{n/b+j}{n/b}} = \frac{1}{(j-1)} \sum_{\mu=0}^{b-1} \frac{1}{\prod_{\nu=1}^{j-1}\left(\nu+\frac{\mu}{b}\right)}$$

since $\alpha_\mu^{b,m}=1$ for $\mu=0,1,\ldots,b-1$, which confirms the result (2.10).

(2) For $m=2$, $\alpha_\mu^{b,2}=(b-|b-1-\mu|)$ for $\mu=0,1,\ldots,2(b-1)$, hence

$$\sum_{n=0}^{\infty} \frac{n+1}{\binom{n/b+j}{n/b}} = j(j-3)! \sum_{\mu=0}^{2(b-1)} \frac{(b-|b-1-\mu|)}{\prod_{\nu=1}^{j-2}\left(\nu+\frac{\mu}{b}\right)}.$$

(3) For $b=2$, $\alpha_{\mu}^{2,m}=\binom{m}{\mu}$ for $\mu=0,1,2,\ldots,m$

$$\sum_{n=0}^{\infty}\frac{\binom{n+m-1}{n}}{\binom{n/2+j}{n/2}}=\frac{j!}{m!\binom{j-1}{j-1-m}}\sum_{\mu=0}^{m}\frac{\binom{m}{\mu}}{\prod_{\nu=1}^{j-m}\left(\nu+\frac{\mu}{2}\right)},$$

$$\sum_{n=0}^{\infty}\frac{\binom{n+m-1}{n}\Gamma\left(\frac{n}{2}+1\right)}{\Gamma\left(\frac{n}{2}+j+1\right)}=\frac{1}{m!\binom{j-1}{j-1-m}}\sum_{\mu=0}^{m}\frac{\binom{m}{\mu}}{\prod_{\nu=1}^{j-m}\left(\nu+\frac{\mu}{2}\right)}.$$

5. Bounding the Series

In the cases when the series (4.1), $S(a,j,m)$ and (4.5), $T(a,j,m)$ cannot be put in explicit closed form, it would be extremely beneficial to be able to bound these sums.

In this section we shall bound the series

$$(j-1)!S(a,j,m)=\frac{1}{j}\sum_{n=0}^{\infty}\frac{\binom{n+m-1}{n}}{\binom{an+j}{an}}.$$

Theorem 5.1. *For $a>0$, $j-1\geq m$ and m a positive real number, let*

$$(j-1)!S(a,j,m)=\frac{1}{j}\sum_{n=0}^{\infty}\frac{\binom{n+m-1}{n}}{\binom{an+j}{an}}, \tag{5.1}$$

then

$$\frac{1}{j}<S(a,j,m)\leq\begin{cases}\frac{B^{1/p}\left(1-mp,\frac{1}{a}\right)}{a^{1/p}(jq-q+1)^{1/q}}\\ \leq\begin{cases}\frac{1}{j-m}; & \text{for } a\geq 1\\ \frac{1}{(j-m)a^{m}}; & \text{for } 0<a<1\end{cases}\end{cases} \tag{5.2}$$

for real numbers p and q where $p>1$, $\frac{1}{p}+\frac{1}{q}=1$ and $B(s,t)$ is the classical Beta function.

Proof. By Hölder's integral inequality [7]

$$(j-1)!S(a,j,m)\leq\left(\int_0^1(1-x)^{(j-1)q}dx\right)^{\frac{1}{q}}\left(\int_0^1\frac{dx}{(1-x^a)^{mp}}\right)^{\frac{1}{p}}.$$

Now,

$$\int_0^1\frac{dx}{(1-x^a)^{mp}}=\frac{1}{a}B\left(1-mp,\frac{1}{a}\right)$$

by the substitution $u=1-x^a$, and hence

$$(j-1)!S(a,j,m)\leq\frac{B^{1/p}\left(1-mp,\frac{1}{a}\right)}{a^{1/p}(jq-q+1)^{1/q}}.$$

The lower bound on $(j-1)!S(a,j,m)$ is $\frac{1}{j}$ since the series (5.1) is one of the positive terms.

The second part of the inequality (5.2) is obtained from

$$\int_{x_0}^{x_1} |f(x)g(x)|\,dx \le \sup_{x\in[x_0,x_1]} |f(x)| \int_{x_0}^{x_1} |g(x)|\,dx.$$

Let $g(x) = (1-x)^{j-1-m}$ and $f(x) = \left(\frac{1-x}{1-x^a}\right)^m$, since $f(x)$ is monotonic on $x \in [0,1]$,

$$\sup_{x\in[x_0,x_1]} \left(\frac{1-x}{1-x^a}\right)^m = \begin{cases} 1 & \text{for } a \ge 1, \\ \frac{1}{a^m} & \text{for } 0 < a < 1, \end{cases}$$

Hence

$$\frac{1}{j} < S(a,j,m) \le \begin{cases} \frac{1}{j-m}, & \text{for } a \ge 1 \\ \frac{1}{(j-m)a^m}, & \text{for } 0 < a < 1, \end{cases}$$

the result (5.2) follows and the theorem is proved. □

The next theorem develops an inequality of (5.1) based on the pre-Grüss inequality [6].

Theorem 5.2.

$$\left|(j-1)!S(a,j,m) - \frac{1}{a(j-m)} \sum_{r=0}^{m} (-1)^r \binom{m}{r} B\left(1-m, \frac{1+r}{a}\right)\right|$$
$$\le \begin{cases} \frac{j-m-1}{2(j-m)\sqrt{2j-2m+1}}, & \textit{for } a \ge 1 \\ \frac{j-m-1}{2a^m(j-m)\sqrt{2j-2m+1}}, & \textit{for } 0 < a < 1, \end{cases}$$

where $B(p,q)$ is the classical Beta function.

Proof. Define

$$V(g,f) = \frac{1}{x_1-x_0}\int_{x_0}^{x_1} f(x)g(x)\,dx - \frac{1}{x_1-x_0}\int_{x_0}^{x_1} f(x)\,dx \cdot \frac{1}{x_1-x_0}\int_{x_0}^{x_1} g(x)\,dx,$$

for $f(x) = \left(\frac{1-x}{1-x^a}\right)^m$, $g(x) = (1-x)^{j-1-m}$ both integrable functions and defined on $x \in [0,1]$, then the pre-Grüss inequality states that

$$|V(g,f)| \le \frac{\Gamma-\gamma}{2}[V(g,g)]^{1/2}$$

for $\gamma \le f(x) \le \Gamma$.

Now for $x \in [0,1]$,

$$\gamma = 0 \le f(x) = \left(\frac{1-x}{1-x^a}\right)^m \le \begin{cases} 1 & \text{for } a > 1, \\ \frac{1}{a^m} & \text{for } 0 < a < 1, \end{cases} = \Gamma$$

and

$$\begin{aligned}
V(g,f) &= \int_0^1 \frac{(1-x)^{j-1}}{(1-x^a)^m} dx - \int_0^1 \left(\frac{1-x}{1-x^a}\right)^m dx \int_0^1 (1-x)^{j-1-m} dx \\
&= (j-1)!S(a,j,m) - \frac{1}{(j-m)a} \sum_{r=0}^{m} (-1)^r \binom{m}{r} \int_0^1 u^{-m} (1-u)^{\frac{1+r}{a}-1} du \\
&= (j-1)!S(a,j,m) - \frac{1}{a(j-m)} \sum_{r=0}^{m} (-1)^r \binom{m}{r} B\left(1-m, \frac{1+r}{a}\right).
\end{aligned}$$

In a similar fashion

$$V(g,g) = \frac{(j-m-1)^2}{(j-m)^2(2j-2m+1)},$$

and

$$[V(g,g)]^{\frac{1}{2}} = \frac{(j-m-1)}{(j-m)\sqrt{2j-2m+1}}.$$

Combining $V(g,f)$ and $[V(g,g)]^{1/2}$, we obtain

$$\left|(j-1)!S(a,j,m) - \frac{1}{a(j-m)} \sum_{r=0}^{m} (-1)^r \binom{m}{r} B\left(1-m, \frac{1+r}{a}\right)\right| \le \begin{cases} \frac{j-m-1}{2(j-m)\sqrt{2j-2m+1}}, & \text{for } a \ge 1 \\ \frac{j-m-1}{2a^m(j-m)\sqrt{2j-2m+1}}, & \text{for } 0 < a < 1. \end{cases}$$

□

Bibliography

[1] R. Apéry, Irrationalitè $\zeta(2)$ and $\zeta(3)$, *Journees Arithmètiques de Luminy, Astérisque,* **61** (1979), 11–13.

[2] N. Batir, Integral representations of some series involving $\binom{2k}{k}^{-1} k^{-n}$ and some related series, *Appl. Math. Comput.*, **147**(3) (2004), 645–667.

[3] N. Batir, On the series $\sum_{k=1}^{\infty} \binom{3k}{k}^{-1} k^{-n} x^k$, *Proc. Indian Acad. Sci.* (Math. Sci.), **115**(4) (2005), 371–381.

[4] T. Mansour and J. West, Avoiding 2-letter signed patterns, Séminaire Lotharingien de Combinatoire, 49, 2002.

[5] T. Mansour, Combinatorial identities and inverse binomial coefficients, *Adv. Appl. Math.*, **28** (2002), 196–202.

[6] M. Matić, J.E. Pečarić and N. Ujević, On new estimation of the remainder in generalised Taylor's formula. *Math. Ineq. and Appl.*, **2** (1999), 343–361.

[7] D.S. Mitrinović, J.E. Pečarić and A.M. Fink, *Classical and New Inequalities in Analysis,* Kluwer Academic Publishers, Boston, 1993.

[8] J. Pla, The sum of inverses of binomial coefficients revisited, *Fibonacci Quart.*, **35** (1997), 342–345.

[9] A.M. Rockett, Sums of the inverses of binomial coefficients, *Fibonacci Quart.*, **19** (1981), 433–437.

[10] A. Sofo, *Computational Techniques for the Summation of Series*, Kluwer Academic/Plenum Publishers, 2003.

[11] B. Sury, Sum of the reciprocals of the binomial coefficients, *European J. Combin.*, **14** (1993), 351–535.

[12] B. Sury, T. Wang and F.Z. Zhao, Identities involving reciprocals of binomial coefficients, *Journal of Integer Sequences*, **7** (2004), Article 04.2.8.

[13] T. Trif, Combinatorial sums and series involving inverses of binomial coefficients, *Fibonacci Quart.*, **38** (2000), 79–84.

[14] S. Weinzierl, Expansion around half-integer values, binomial sums and inverse binomial sums, *J. Math. Physics,* **45**(7) (2004), 2656–2673.

[15] A.D. Wheelon, On the summation of infinite series in closed form, *Journal of Applied Physics*, **25** (1954), 113–118.

[16] F. Zhao and T. Wang, Some results for sums of the inverse of binomial coefficients, *Integers: Electronic Journal of Combinatorial Number Theory*, **5**(1) (2005), #22.

Index